Sustainable Tourism in
Southern Africa

PEFC™

PEFC/16-33-111
CATG-PEFC-052
www.pefc.org

ASPECTS OF TOURISM
Series Editors: Chris Cooper, *Nottingham University Business School, UK*;
C. Michael Hall, *University of Canterbury, New Zealand* and Dallen J. Timothy,
Arizona State University, USA

Aspects of Tourism is an innovative, multi-faceted series, which comprises
authoritative reference handbooks on global tourism regions, research volumes,
texts and monographs. It is designed to provide readers with the latest thinking
on tourism worldwide and push back the frontiers of tourism knowledge. The
volumes are authoritative, readable and user-friendly, providing accessible sources
for further research. Books in the series are commissioned to probe the relationship
between tourism and cognate subject areas such as strategy, development, retailing,
sport and environmental studies.

Full details of all the books in this series and of all our other publications can be
found on http://www.channelviewpublications.com, or by writing to Channel
View Publications, St Nicholas House, 31–34 High Street, Bristol BS1 2AW, UK.

ASPECTS OF TOURISM
Series Editors: Chris Cooper, C. Michael Hall and Dallen J. Timothy

Sustainable Tourism in Southern Africa
Local Communities and Natural Resources in Transition

Edited by
Jarkko Saarinen, Fritz Becker,
Haretsebe Manwa and Deon Wilson

CHANNEL VIEW PUBLICATIONS
Bristol • Buffalo • Toronto

Library of Congress Cataloging in Publication Data
A catalog record for this book is available from the Library of Congress.
Sustainable Tourism in Southern Africa: Local communities and
Natural Resources in Transition
Edited by Jarkko Saarinen et al.
Includes bibliographical references and index.
1. Tourism--South Africa. 2. Environmental protection--South Africa.
I. Saarinen, Jarkko II. Title.
G155.S57S87 2009
338.4'79168–dc22 2009017289

British Library Cataloguing in Publication Data
A catalogue entry for this book is available from the British Library.

ISBN-13: 978-1-84541-109-1 (hbk)
ISBN-13: 978-1-84541-108-4 (pbk)

Channel View Publications
UK: St Nicholas House, 31–34 High Street, Bristol BS1 2AW, UK.
USA: UTP, 2250 Military Road, Tonawanda, NY 14150, USA.
Canada: UTP, 5201 Dufferin Street, North York, Ontario M3H 5T8, Canada.

The policy of Multilingual Matters/Channel View Publications is to use papers
that are natural, renewable and recyclable products, made from wood grown in
sustainable forests. In the manufacturing process of our books, and to further
support our policy, preference is given to printers that have FSC and PEFC Chain
of Custody certification. The FSC and/or PEFC logos will appear on those books
where full certification has been granted to the printer concerned.

Typeset by Techset Composition Ltd., Salisbury, UK.
Printed and bound in Great Britain by the MPG Books Group.

Contents

Cases and Issues vii
Figures ix
Tables xi
Plates. xiii
Abbreviations xv
Contributors xix
Preface xxi

Part 1: Introduction and Contexts

1 Introduction: Call for Sustainability   3
   Jarkko Saarinen, Fritz Becker, Haretsebe Manwa and Deon Wilson

2 Tourism Development in Southern Africa: Patterns,
   Issues and Constraints 20
   Christian M. Rogerson

3 Tourism Policy and Politics in Southern Africa. 42
   C. Michael Hall

4 Women and Tourism in Southern Africa 61
   Haretsebe Manwa

5 Sustainable Tourism: Perspectives to
   Sustainability in Tourism 77
   Jarkko Saarinen

Part 2: Tourism Development and Local Policies of Sustainability

6 Tourism, Conservation Areas and Local Development
   in Namibia: Spatial Perspectives of Private and
   Public Sector Reform. 93
   Fritz Becker

7 Commercialization of National Parks: South Africa's
   Kruger National Park as an Example 116
   David M. Mabunda and Deon Wilson

v

8   Natural Resource-based Tourism and
    Wildlife Policies in Botswana................................. 134
    Julius Atlhopheng and Kutlwano Mulale

9   Tourism, Nature Conservation and Environmental
    Legislation in Namibia 150
    Susanne Scholz

10 Transfrontier Conservation and Local Communities............ 169
    Maano Ramutsindela

**Part 3: Tourism, Local Communities and
Natural Resources in Transition**

11 Village-based Tourism and Community Participation: A Case
    Study of the Matsheng Villages in Southwest Botswana......... 189
    Naomi Moswete, Brijesh Thapa and Gary Lacey

12 The Socio-economic Impacts of Tourism in the
    Okavango Delta, Botswana 210
    Joseph E. Mbaiwa and Michael B.K. Darkoh

13 Sustainable Tourism on Commonages: An Alternative to
    Traditional Agricultural-based Land Reform in
    Namaqualand, South Africa................................. 231
    Sharmla Govender-van Wyk and Deon Wilson

14 Local Food as a Key Element of Sustainable Tourism
    Competitiveness... 253
    Gerrie E. du Rand and Ernie Heath

15 Conclusions and Critical Issues in Tourism and
    Sustainability in Southern Africa.......................... 269
    Jarkko Saarinen

Index 288

Cases and Issues

2.1 Tourism in the Victoria Falls: A Zimbabwean Perspective 23

2.2 Tourism Benefiting the Poor: PPT 37

3.1 Botswana Tourism Policy: Diversifying 'Low-Volume–
High-Value' Development Strategy.......................... 44

4.1 Tourism, Gender and the Exotic 'Other': The Umhlanga
Ceremony (Reed Dance) in Swaziland 64

5.1 The Nata Sanctuary Community-based Project................ 83

6.1 Cultural Heritage Tourism and Sustainability in
South Africa: The Case of the Cradle of Humankind.......... 101

7.1 Tourist Satisfaction in the Etosha National Park in 2008 121

8.1 Recreational Quad-biking on the Coastal Dunes between
Swakopmund and Walvis Bay, Namibia 144

9.1 The Namibia Community-based Tourism Assistance Trust 161

10.1 'When Cultural Values are Marginalised': Impacts of
the GLTP to Local People in Mozambique................... 179

11.1 Communal Areas Management Programme for Indigenous
Resources in Zimbabwe (CAMPFIRE) 190

13.1 Community Perceptions and Participation in the
Management of the Isamangoliso Wetland Park,
South Africa: Towards a Sustainable Tourism
Development Initiative 233

15.1 Global Sustainable Tourism Criteria 272

15.2 United Nations' Millennium Project Goals 275

Figures

1.1 Location map of southern African countries and selected
    case study places and regions. 15

2.1 Location map of southern African region,
    including major tourism attractions 22

2.2 The promotion of regional tourism from Botswana to
    South Africa and Mozambique 29

2.3 International tourism arrivals by country, 2005. 30

2.4 Receipts from international tourism 30

2.5 Source of international tourism arrivals in
    South Africa, 2005. 33

3.1 Tourism policy fields 49

3.2 Multi-layered tourism governance 53

4.1 Inter-related factors to empower women 73

5.1 Tourism area life cycle model. 81

6.1 Spatial classification for regional planning in
    Namibia (example). 109

6.2 Livelihood outcomes from tourism and
    improvement through CBT. 111

7.1 Map of South Africa showing the Kruger
    National Park 117

7.2 Percentage of responses of South Africans and foreigners
    regarding the outsourcing of accommodation. 124

7.3 Percentage of responses on price increase and
    continued visits to the park 125

7.4 Percentage of responses about different accommodation
    rates for foreigners and South Africans 125

7.5 Percentage of responses regarding an appropriate
rate of increase for foreigners 126

7.6 Percentage of responses to different levels of satisfaction
with accommodation 126

7.7 Percentage of responses to different levels of satisfaction with
accommodation of the South African and foreign samples 127

7.8 Responses of South Africans and foreigners regarding
different accommodation rates for foreigners. The respondent
who did not specify country of origin is not included in the
results below 129

8.1 Various land uses and their proportions in Botswana 135

9.1 Conservancies and protected areas 159

12.1 Map of Botswana showing the Okavango Delta 211

12.2 Interpretation of sustainable tourism 212

13.1 Map of selected commonage projects 239

14.1 Destination competitiveness framework 256

14.2 Inter-relatedness between sustainable competitiveness,
food tourism and destination marketing management 258

14.3 Relationship between identified gaps, actions and
recommended strategies 262

Tables

1.1   General elements of success in CMT destination
development 5

1.2   UNWTO's (before 2003 as WTO) official statements
and declarations related to sustainable development
of tourism 10

1.3   International tourist arrivals (in millions) and the
relative share (%) of arriving tourists by UNWTO regions.
UNWTO's sub-regions do not include southern Africa 12

1.4   Some of the key characteristics of the selected southern
African countries 14

2.1   Occupancies by region 2005 compared with 2004. 24

2.2   International tourism arrivals by country 1990–2005 (1000s) 31

2.3   International tourism receipts by country 1990–2005
(US$ million) 31

2.4   Travel advisories for British citizens travelling to
southern Africa, January 2008 36

3.1   International actors in tourism-related donor assistance
in Mozambique. 54

4.1   RETOSA Board (2005–2006) by gender and sector 69

6.1   Categories of protected areas in Namibia (proposed) 96

6.2   Conservancies in communal areas in Namibia (2006) 108

6.3   Conservancy income (2006) 111

7.1   Tourists' satisfaction with the services and opportunities
in the Etosha National Park (%) 123

7.2   Respondents from according to origin 124

8.1 Landforms and their significance in Botswana 140

8.2 Summary of some major landscapes/landforms attractions
in Botswana. 144

9.1 Strategic framework (1990–2007): Tourism and nature
conservation in Namibia (selected documents). 152

10.1 Tourism potential and arrivals in TFCAs 177

11.1 Tourism-related attractions. 200

12.1 Road network linking Okavango region 213

12.2 Employment at Sankoyo, Khwai and Mababe, 1997–2007. 217

12.3 Revenue generated by Sankoyo, Khwai and Mababe,
1997–2007. 218

12.4 Number of tourists and revenue collected at Moremi Game
Reserve, 1998–2007. 220

13.1 Comparing the main components of sustainable
development with current land redistribution policy
and implementation. Table shows only some factors crucial
to land reform 239

13.2 Livestock farming on commonages. 241

13.3 Advantages and disadvantages of livestock farming
on commonages (N = 34). 242

13.4 Improvement/non-improvement of quality of life (N = 38) 242

13.5 Training received 247

13.6 Opinions on the plans for the conservancy 247

13.7 Economic and social spin-offs from conservancy project. 248

Plates

4.1 Maidens are carrying the reed to the Queen Mother at the
     Umhlanga (Reed Dance) ceremony in Swaziland. 66

5.1 The entrance of the Nata Sanctuary,
     Makgadikgadi Salt Pans. In October 2008, the main
     parts of the Sanctuary along the nearby private
     Nata Lodge were burned by extensive bush fires. 84

11.1 Entrance to KD1 (hunting zone) Wildlife
      Management Area 196

11.2 Nqwaa Khobee Xeya Trust CBO guesthouse in Ukhwi 201

11.3 Nqwaa Khobee Xeya Trust CBO campsite in Ngwatle. 202

13.1 Nanasan, Spiönkop and Taaibosmond give an indication
      of arid conditions on the commonage farms 243

Abbreviations

ANC	African National Congress (RSA)
ASATA	Association of South African Travel Agents (RSA)
BNES	Botswana National Ecotourism Strategy
BSA	backpacking South Africa
BABASA	Bed and Breakfast Association of South Africa
BARSA	Board of Airline Representatives of South Africa
BNES	Botswana National Ecotourism Strategy
CAMPFIRE	Communal Areas Management Programme for Indigenous Resources
CBD	convention on biological diversity
CBNRM	Community-based Natural Resource Management
CBO	community-based organisation
CBT	community-based tourism
CBTE	community-based tourism enterprise
CHA	controlled hunting areas (BW)
CMT	conventional mass tourism
COASA	Coach Operators Association of Southern Africa
COP 7	Seventh Conference of Parties
COSATU	Congress of South African Trade Unions
CPA	Communal Property Association (RSA)
DA	Democratic Alliance (RSA)
DDP	District Development Plan (BW)
DFID	Department for International Development (UK)
DLA	Department of Land Affairs (RSA)
DMO	destination management organisation
DTI	Department of Trade and Industry (RSA)
DWNP	Department of Wildlife and National Parks (BW)
EIA	environment impact assessment
FEDHASA	Federated Hospitality Association of South Africa
FGASA	Field Guides Association of Southern Africa
GHASA	Guest House Association of Southern Africa
GDP	gross domestic product
GLTP	Great Limpopo Transfrontier Park
GSTC	Global Sustainable Tourism Criteria

IFP	Inkatha Freedom Party (RSA)
IRDNC	Integrated Rural Development and Nature Conservation (NAM)
KNP	Kruger National Park (RSA)
KTP	Kgalagadi Transfrontier Park
KZN	KwaZulu-Natal (Province of RSA)
LDP1	Livestock Development Project (BW)
LHWP	Lesotho Highland Water Project
LRAD	Land Redistribution for Agricultural Development (RSA)
MET	Ministry of Environment and Tourism (NAM)
MFDP	Ministry of Finance and Development Planning (BW)
MDGs	Millennium Development Goals
MLG	Ministry of Local Government and Land (BW)
NACOBTA	Namibia Community-based Tourism Assistance Trust
NACSOs	Namibian Association of Community-based Support Organisations
NDP2	National Development Plan II (NAM)
NDP9	National Development Plan 9 (BW)
NEPAD	new partnership for Africa's development
NGO	non-governmental organisation
NKXT	Nqwaa Khobee Xeya Trust (BW)
NPB	National Parks Board (RSA)
ODI	Overseas Development Institute (UK)
OECD	Organisation for Economic Co-operation and Development
OPT	Okavango Poler's Trust (BW)
PPP	Public–private partnership
PPT	pro-poor tourism
PWMA	Parks and Wildlife Management Act (NAM)
RAD	remote area dweller (BW)
RASA	Restaurant Association of South Africa
RETOSA	Regional Tourism Organization of Southern Africa
RDC	Rural District Council (ZIM)
RISDP	Regional Indicative Strategic Development Plan
SAACI	Southern African Association for the Conference Industry
SACP	South African Communist Party
SADC	Southern African Development Community
SANParks	South Africa National Parks
SAR&H	South African Railways and Harbours
SATSA	Southern Africa Tourism Services Association
SAVRALA	Southern African Vehicle Rental and Leasing Association

SDI	spatial development initiative
SME	small and medium-sized enterprise
SRDC	Southern Africa Research and Documentation Centre
TBCSA	The Tourism Business Council of South Africa
TFCA	Transfrontier Conservation Area
TGLP	Tribal Grazing Land Policy (BW)
THETA	Tourism, Hospitality, Education and Sports Training Authority
TWN	Third World Network
UNCED	United Nations Conference on Environment and Development
UNEP	United Nations Environment Programme
UNESCO	United Nations Educational, Scientific and Cultural Organisation
UNWTO	United Nations World Tourism Organization
VFR	visiting friends and relatives
WIDSAA	Women in Development Southern Africa Awareness
WMAs	wildlife management areas
WTO	World Tourism Organization (before UNWTO)
WTTC	World Travel & Tourism Council
WWF-LIFE	World Wildlife Fund – Living in a Finite Environment
ZTA	Zimbabwe Tourism Authority

Contributors

Julius Atlhopheng, Department of Environmental Science, University of Botswana, Private Bag 704, Gaborone, Botswana (atlhophe@mopipi.ub.bw).

Fritz Becker, Department of Geography, History and Environmental Studies, University of Namibia, Private Bag 13301, Windhoek, Namibia (fobecker@unam.na; fobecker@mweb.com.na).

Michael B.K. Darkoh, Department of Environmental Science, University of Botswana, Private Bag 704, Gaborone, Botswana (darkohmb@mopipi.ub.bw)

Felicite A. Fairer-Wessels, Department of Tourism Management, Cnr Duxbury Road and Hilda Street, University of Pretoria, Pretoria, 0002, South Africa (felicite.fairer-wessels@up.ac.za).

Sharmla Govender-van Wyk, Directorate: Redistribution Implementation Systems at the Department of Land Affairs, Private Bag X833, Pretoria 0001, South Africa (sgovendervanwyk@dla.gov.za).

C. Michael Hall, Department of Management, University of Canterbury, Private Bag 4800, Christchurch, New Zealand 8140 (michael.hall@canterbury.ac.nz)

Ernie Heath, Department of Tourism Management, Cnr Duxbury Road and Hilda Street, University of Pretoria, Pretoria, 0002, South Africa (ernie.heath@up.ac.za)

Gary Lacey, Tourism Research Unit, Monash University, Melbourne, PO Box 1071, Narre Warren VIC 3805, Australia (Garylacey1@hotmail.com)

David Mabunda, South African National Parks (SANParks), PO Box 787, Pretoria 0001, South Africa (davidm@sanparks.org)

Aderito Machava, Department of History, Faculty of Arts and Social Sciences, Eduardo Mondlane University, PO Box 257, Maputo–Mozambique (aderito.machava@uem.mz, admachava@yahoo.com)

Lindisizwe M. Magi, Centre for Recreation and Tourism, University of Zululand, 34 Kinmont Crescent, Carrington Heights, Durban, 4001, South Africa (lmmagi@pan.uzulu.ac.za; lmmagi@iafrica.com)

Haretsebe Manwa, Department of Management, University of Botswana, Private Bag UB 00701, Gaborone, Botswana (manwah@mopipi.ub.bw; hmanwa@yahoo.com.au)

Kenneth Matengu, Multidisciplinary Research Centre, University of Namibia, Private Bag 13301, Windhoek, Namibia (kmatengu@unam.na)

Joseph E. Mbaiwa, University of Botswana, Harry Oppenheimer Okavango Research Centre, Private Bag 285, Maun, Botswana (mbaiwaj@mopipi.ub.bw)

Naomi Moswete, Department of Environmental Science, University of Botswana, Private Bag 704, Gaborone, Botswana (nomsamoatshe@yahoo.com)

Kutlwano Mulale, Department of Environmental Science, University of Botswana, Private Bag 704, Gaborone, Botswana (mulalek@mopipi.ub.bw)

Thandi A. Nzama, Centre for Recreation and Tourism, University of Zululand, Private Bag X1001, KwaDlangezwa, South Africa (atnzama@pan.uzulu.ac.za)

Gerrie du Rand, Department of Consumer Science, Cnr Duxbury Road and Hilda Street, University of Pretoria, Pretoria, 0001, South Africa (gerrie.durand@up.ac.za)

Christian M. Rogerson, School of Geography and Environmental Studies, University of the Witwatersrand, Private Bag 3, PO WITS 2050, Johannesburg, South Africa (Christian.Rogerson@wits.ac.za)

Jarkko Saarinen, Department of Geography, PO Box 3000, University of Oulu, FI-90014, Finland; and Department of Management, University of Botswana, Private Bag UB 00701, Gaborone, Botswana (jarkko.saarinen@oulu.fi; jarkko.saarinen@mopipi.ub.bw)

Susanne Scholz, Department of Geography, History and Environmental Studies, University of Namibia, Private Bag 13301, Windhoek, Namibia (sscholz@unam.na)

Brijesh Thapa, Department of Tourism, Recreation and Sport Management, Center for Tourism, Research and Development, University of Florida, Gainesville, FL 32611-8208, USA (bthapa@hhp.ufl.edu)

Deon Wilson, Department of Tourism Management, Cnr Duxbury Road and Hilda Street, University of Pretoria, Pretoria, 0002, South Africa (deon.wilson@up.ac.za)

Preface

Tourism is a very important economic and social element in southern Africa. Although tourism and its developmental nature are always polarised, the impacts of tourism are relatively widely spread in the region. New places are constantly introduced to tourism, and the existing destinations and attractions are modified and further developed in order to satisfy the needs of growing numbers of international as well as evolving domestic tourists. As a result, tourism has become a major economic driver in the region with broad scale of benefits ranging from additional incomes and employment, improved public infrastructures, facilities and services, new incoming investments and increased possibilities for entrepreneurship and more international business environments and contacts, for example. Tourism has also a wide range of other kinds of evident or potential impacts on environment, economy, people and their everyday life and cultures. This has highlighted the role of sustainability in tourism. The issue of sustainable development and thinking the impacts and changes proactively are important in southern Africa, where the attractiveness of natural and traditional cultural environments, which are sensitive to the changes caused by poorly managed and controlled tourism, plays the key role in tourism and its future prospects in the region.

With the increasing role of the tourism industry in southern Africa, the need to research and study tourism, its developmental nature and impacts has been emphasised recently. During the past few years, the research input has increased, and also educational programmes and courses have evolved in tourism and related fields. In the region, the first dedicated Department of Tourism Management was established at the University of Pretoria, South Africa, in 1997. Most probably the youngest university level unit, the Department of Tourism and Hospitality Management, was established at the University of Botswana (UB) in 2009. Thus, the focus on tourism studies is growing, but as in many other regions in the world, the elements and processes of sustainable development in tourism development, policies and education still call for more work and interests in the academia and beyond. The aim of this book is to contribute to the academic field of tourism studies by providing an introduction to the connections between tourism, sustainability and development with a range of case studies and examples from the region. The volume has an emphasis on policy issues in sustainable tourism and as such it aims to be used by specialists, academics and students of southern African and international tourism, management, environmental studies, geography and social sciences. The main focus of the book is on tourism, sustainability and natural resource-dependent communities facing processes of transition. These processes are generally long and gradual with direct and/or indirect links to evolving tourism. Although the book is recognising the key role of tourism in transition, the social, cultural, economic and political contexts of tourism and local communities are also highlighted. The academic value of the contributions has been enhanced by a peer review process.

A number of individuals and organisations deserve a word of thanks. The home universities of editors have been supportive in the collaboration. The Finnish Foreign Ministry's financial support through the Institutional Collaboration Instrument (ICI) and the project *Sustainable Tourism in Southern Africa* has made it possible for the editors and contributors to meet and discuss about the volume in hand, and on related course and collaborative plans in future. In addition, the North-South-South Higher Education Programme coordinated by Centre for International Mobility (CIMO, Finland) has provided a very good platform to meet and develop sustainable tourism-related courses and course materials, based on the book project between the universities of Botswana, Cape Town, Namibia, Oulu and Pretoria. We would like to thank Elinor Robertson and the Channel View staff and C. Michael Hall (Series Editor) for their interest, support and great help with the book, starting from an early stage of the project till the printing of the finished product in hand.

Jarkko Saarinen would like to thank his co-editors for their efforts in working on the edited book and for the whole process and opportunities to discuss and spend time with them in their home institutions and in Oulu, close to the Arctic Circle. The final part of the book was edited while Jarkko was working at the UB as a Visiting Professor – sincere thanks to the UB, the Faculty of Business and Haretsebe for hosting him. Jarkko would also like to thank colleagues at the Department of Geography 'back home' and especially the Tourism Geographies Research Group at the department. Moreover, thanks are due to Richard Butler, Rene van der Duim, David Duval, David Fennell, Willem Ferguson, Sanette Ferreira, Antti Honkanen, Petri Hottola, Leon Hugo, Jari Järviluoma, Alan Lew, Dorothea Meyer, Dieter Müller, Anja Mäläskä, Maaria Niskala, Kaija Pajala, Dallen Timothy, Regina Scheyvens and Gustav Visser – among many others – for discussions, collaboration and support, and the OYUS Rugby Club for relaxing time, when no academic matter weighs a thing. Final thanks are for Satu, Mira and Katlego waiting at home or travelling with.

Fritz Becker, who enjoyed Jarkko's vigour in propelling the project and the spirit of the editing team, has pleasure in expressing a word of thanks for all the support the Namibian contributors received from Dr Martin Hipondoka and his Laboratory for Spatial Analysis of the Department of Geography, History and Environmental Studies at the University of Namibia; his professional assistance in the thematic map work is always appreciated. On a personal note, Fritz is indebted to Ria and Franka who patiently waited at home and seldom had the opportunity to join travelling for the project. Haretsebe Manwa would like to express her appreciation and thanks to Professor Chanda who nominated her to represent the Department of Environmental Science on the Book project; Professor Darkoh, Professor Philip Pearce and Professor Gianna Moscardo for being my inspiration. The most important person in my life: Joe, whose support has always made the long journey seem so simple and enjoyable. Deon Wilson would like to thank Jarkko Saarinen for initiating this project, acquiring funding and providing support. Your professional leadership as editor of this book is much appreciated. A special word of thanks to the co-editors.

Further, the individual contributors of the book deserve our greatest thanks – without them the book would not exist. We trust that this book will be informative and a source of inspiration to all readers.

Jarkko Saarinen, Fritz Becker,
Haretsebe Manwa and Deon Wilson
Gaborone, Windhoek and Pretoria

Part 1

Introduction and Contexts

Chapter 1
Introduction: Call for Sustainability

JARKKO SAARINEN, FRITZ BECKER, HARETSEBE MANWA
and DEON WILSON

Introduction

Tourism has become a global industry with increasing impacts on environment, regional and local development. In many southern African countries, tourism provides new opportunities, jobs and economic benefits for local communities. Currently, many countries in the region perceive tourism promotion as a suitable and relatively inexpensive strategy that can be used to attract foreign direct investment (Binns & Nel, 2002). As a result of growing tourism activities, many places and rural areas in the region are tied to the global industry and related cultural, social, economic and political networks. Thus, tourism has become an important policy tool for community and regional development in southern Africa (Rogerson, 1997). Tourism also has a significant potential to influence and change the use of natural and cultural resources in the region. This potential has highlighted the role of sustainability in tourism development and has turned tourism into not only an economic but also a social and political agent that affects a wider natural and socio-cultural environment in various ways; for example, sustainable utilisation of natural resources and improved quality of life of the communities living adjacent to tourism resources.

The attractiveness and recent success of southern Africa in tourism development has been based on its diversity. The White Paper of Development and Promotion of Tourism in South Africa (Government of South Africa, 1996), for example, states that the attractiveness of the region is based on relatively accessible wildlife, beautiful scenery, unspoiled nature, diverse traditional and township cultures and pleasant climate (see also Ministry of Environment and Tourism, 2005; Rogerson & Visser, 2004). Similarly, other countries in the region base their strengths on the natural environment and, increasingly, on cultural settings (Saarinen, 2007; Van Veuren, 2001; World Travel and Tourism Council, 2007).

However, many of these elements of attraction are sensitive to the changes caused by uncontrolled tourism development and management, which are subject to a multitude of other wider processes such as global climate change and political stability of the region (see Hall, Chapter 3, this volume). All these components and factors call for a deeper understanding of tourism, its sustainability and the need to develop sustainable tourism strategies, knowledge based on higher education and academic research efforts in the region.

At policy level, tourism is currently viewed as an essential sector for regional and national reconstruction and development in the tourism policy of the southern Africa Development Community (SADC). In this sense, the rationale of tourism development has evolved towards the idea of tourism as a tool for local and regional economic development (see Rogerson, 1997), and, recently, a relatively new kind of idea of tourism as an instrument of social and economic empowerment is also taking centre stage (Binns & Nel, 2002; Scheyvens, 2002). In this respect, there are currently many regional and local development programmes highlighting the significance of tourism in both regional development and black empowerment. The New Partnership for Africa's Development (see Goudie *et al.*, 1999) underlines this aspect and stresses the potential role of tourism in social and economic development. The background of these new policy needs is the growing role of tourism in the region and concerns over the local-level benefits of the increasingly global industry; tourism is accepted as a vital export industry, but which must benefit the local residents of the places that tourists actually visit.

Tourism's role is also seen positively in global contexts such as the United Nation's Millennium Project and its goals and targets (see Telfer & Sharpley, 2007). Based on this change of orientation, tourism could, if not should, be more often used as a development tool for poverty reduction, ensuring environmental sustainability, developing a global partnership for development and empowerment of previously neglected communities and social groups, for example. Especially the issue of sustainable tourism acknowledging community-based tourism (CBT) and pro-poor tourism (PPT) has been given a central position in these development goals and discussions.

Changing Tourism and Impacts

The scale, increasing role and diversity of global tourism have resulted in growing environmental impacts. Also in southern Africa, tourism has resulted in numerous positive economic, environmental and social impacts. Although tourism has a tremendous capacity for generating development in destination regions, the growing impacts of tourism may also lead to a range of evident and potential problems as well as of environmental, social, cultural, economic and political issues in tourism destinations and systems, creating a need for alternative and more environment and host-friendly practices in tourism development, planning and policies (Hall, 2000). In general, these are labelled under the concept of sustainable tourism. Sustainable tourism, its practices and code of conducts are often seen as highly beneficial for local communities, their well-being, natural environment and local development (Saarinen, Chapters 5 and 15 in this book). In contrast to conventional mass tourism (CMT) activities, sustainable modes of tourism are regarded as more locally oriented, and culturally and socially responsible.

However, the past success of tourism has not been influenced much by ideas related to sustainability. Table 1.1 aims to demonstrate key elements of CMT and factors influencing the success of tourist destination development. In relation to a certain specific context, there may be some other crucial elements involved, which are not listed here and, for example, the issues of governance, legislation and policy are taken for granted in this generalised model. Generally, in tourism development, discussions on the accessibility of a destination have often been often highlighted (see Hall, 2005; Rogerson, Chapter 2 in this book). Accessibility is a very important factor that explains the colossal public investments in roads and, particularly, airport facilities in mass tourism regions, for example. However, accessibility alone does not make tourism happen. In order to have tourism and tourists, people need to be motivated to travel. Tourist motivation relates to certain general pull factors and specific attractions that a destination can provide. Safety and security have always played an important role in mass tourism, but nowadays it is an inevitable aspect among other elements of the attractiveness of a destination. In addition, to capitalise, that is, turn good accessibility and attractiveness into real tourism business and tourist flows, there need to be basic services for visitors to use.

The primary factors are followed by marketing, highly specialised knowledge in business skills and the nature of the basic physical infrastructure. These so-called secondary elements are partly integrated and

Table 1.1 General elements of success in CMT destination development

Primary elements	Secondary elements	Other elements
Accessibility	Marketing	Good personnel skills
Attractions, including safety	Knowledge on business economy and skills	Research information
Basic services	Basic infrastructure and low wage level	Links to local economy and communities

Source: Modified from Hall (2005)

dependent on the primary elements and vice versa. Although marketing, as such, is also an important element in CMT or any other destination development, it still represents a secondary factor; without accessibility, attractions and basic services there is nothing to be marketed. An entry point to the tourism business has traditionally been relatively low (at least compared with many other large industries) without sound knowledge of or education in the field (Hall, 2005). There has also been more emphasis on quantity and quality in CMT resort infrastructure and development. In addition to primary and secondary elements, there are a number of other factors, such as personnel skills, accessibility to research information and a level of integration into local economies and communities, potentially affecting tourism development. However, these have usually played a relatively minor role in CMT development.

Currently, there are several ongoing processes of change that are challenging the CMT development model and priorities (see Holden, 2007). On a general level, the internationalisation and rapid globalisation of the tourism industry has changed or is changing the markets (see Gössling & Hall, 2005). Although globalisation has spread the CMT developments to new areas with increasing impacts, it has also diversified the markets and brought new kinds of tourist segments and elements of demand to new destinations. The emergence of the more individual, international and potentially responsible tourist and the decrease in the importance of mass tourism have been highlighted in the literature since the mid-1990s (see Poon, 1993). However, there is no clear empirical evidence yet that the core of the industry, representing CMT, is factually declining globally. On the other hand, there is an increasing tourism demand referring to this so-called _new tourism_ that is largely based on the debated assumption of the crisis in modern tourism, specifically the mass tourism industry.

According to Poon (1993), changes in technology, production and consumption modes, management strategies and socio-economic contexts will evidently lead to the end of conventional tourism and towards alternative forms of tourism. As said, this 'end of mass tourism' discussion can be critically questioned, but there is an evident background process of change based on wider shifts in consumption and production modes in Western countries. That change in the literature is described as moving from Fordist production towards post-Fordist production and the related new ways of consuming (Urry, 1990: 14, see Beck & Beck-Gernsheim, 2001; Williams & Shaw, 1998), the trend of which is also affecting the supply and demand patterns of the southern Africa tourism scene. It represents a transformation towards more individually oriented production, marketing and consumption with deeper preferences for small-scale and high-class products. These developments have resulted in a magnitude of new lodges, game parks and other elements referring to a commercialisation of nature and local cultures in the region (see Govender van Wyk & Wilson,

Chapter 13 in this book). In this respect, these new tourists are considered fundamentally different from the 'old' conventional ones; they are expected to be more quality and environmentally conscious, flexible, independent, experienced and community-centred (see Duffy, 2002; Fennell, 1999). In a way, the shift is a classical matter of distinction between a tourist and a traveller; that is, travellers are often regarded as more knowledgeable, educated, responsible, adventurous and individual compared to tourists moving in crowds looking for a familiar, safe and touristic *sun–sea–sand* kind of environments.

In addition to globalisation with new demand and supply patterns and the emergence of the new tourist, the growing need for sustainability has challenged the CMT development model. Sustainable tourism, which will be discussed in detail later (Saarinen, Chapter 5 in this book), has also initiated several alternative modes of tourism, such as CBT and PPT, that also have their connections to the premises of the new tourism. These alternative forms of tourism are partly responses to the globalisation of the industry, its new scale and nature of its impacts. In this respect, all these larger changes highlight the new role of the environment and local people in tourism development with needs to integrate tourism to the local economy and regional development (see Moswete *et al.*, Chapter 11 in this book). In addition, greater expertise to meet the needs of new demand patterns and the call for a research-based knowledge of the tourism impacts and their management are increasingly needed. Although these changes and factors may probably not replace the traditional primary elements of CMT, they challenge the marginal positions of the 'other elements' also in CMT development (see Table 1.1). Especially the need for sustainability has been the driving force in this process of change and the contextualisation of the industry with its social and physical environments.

Sustainability and Tourism

Sustainable development

The demand for more environmentally sensitive practices in tourism grew rapidly in the 1980s. During the 1990s, the issue of sustainability became an idea and platform that started to guide the economic and political structures of the whole tourism system and its development (Bramwell & Lane, 1993; Mowforth & Munt, 1998; Pigram, 1990). The term and idea of sustainability was transferred to tourism from the ideology of sustainable development, following the publication of the Brundtland Commission's report *Our Common Future* in 1987 (WCED, 1987).

The commission's report defines sustainable development as a process that meets the needs of present generations without endangering the ability of future ones to meet their own needs (WCED, 1987). According to the report, sustainability rests on three integrated elements – the ecological, socio-cultural and economic – and there are three fundamental principles in sustainable development: futurity, equity and holism (Redcliffe & Woodgate, 1997). In short, futurity refers to the needs of future generations, that is, a long-term perspective for evaluating the impacts of human activities and socio-economic development, the demand for 'equity' states that different generations should have fair and equal opportunities, and that this should apply to all people, present and future (inter- and intragenerational equity), and the holistic aspect ('holism') implies that development should be considered within broad (global, political, social, economic, and ecological) contexts and perspectives – not only on a local, for example, destination scale. In combination, futurity, equity and holism form the concept of *geographical equity*, in which present and future generations are not disproportionately disadvantaged on the grounds of location in space (see Hunter, 1995).

From the UN 'Earth Summit' in 1992, the need to enforce the principles of sustainable development within wider economic and social development processes highlighted the role of sustainability in tourism and its potential for advancing the goals of sustainable development (Berry & Ladkin, 1997; Butler, 1990, 1991; Hall, 1998). The growing need for sustainability in tourism was also a result of increased knowledge and concern regarding the impacts of tourism and environmental issues in general (see Holden, 2003: 95–96). Many of these concerns date back to the 1960s and 1970s (see Gössling & Hall, 2005; Hall & Lew, 1998; Mathieson & Wall, 1982), reflecting the discussions and concerns over the impacts of economic and population development as well as the limits to growth (see Meadows *et al.*, 1972). In addition, the North/South divide became evident in the environmental debate at that time and was mirrored in tourism development discussions (see Britton, 1982, 1991; Turner & Ash, 1975).

Although these concerns on the limits to growth were truly global in scale, in tourism they were channelled to destination-level analysis on the impacts of tourism and how to define the limits of growth and prevent the negative results of development in tourism destinations (see Gössling & Hall, 2005; Inskeep, 1991). Rather than stating 'the ultimate limits to growth', questions were more concerned with issues and processes limiting the growth and industry's future, that is, limits of growth in tourism. However, the message is the same: the negative outcome (collapse) was not inevitable if tourism development actors change their policies and practices (Saarinen, 2006).

Sustainable tourism and the limits of growth

In tourism, the concept of sustainable development has emerged as a new paradigm (Holden, 2003; Macbeth, 2005). The definition of the concept has been described as complex, normative, imprecise and non-operational

(Hughes, 1995; Liu, 2003; Sharpley, 2000). However, it is not only the obvious vagueness of the WCED's (1987) suggestion or multiple later definitions (see Elliott, 1994: 71), but also the conflict of interests, which cause the fuzzy picture of what sustainability is all about (Cater, 1993; Duffy, 2002; Wall, 1997). The concept of sustainable development is ideologically and politically contested, and it needs to cover a broad range of interests, which have no easily identifiable common denominator (Spangenberg, 2005). Sustainable development is an anthropocentric approach and it embraces the very contradictory ideas in tourism. In general, sustainable development implies that economic growth is needed and acceptable and that the benefits of such development should be available for all. On the other hand, it argues that economic growth causes environmental problems, which is damaging to all (Milne, 1998; Redcliffe, 1987). While sustainable development is a problematic concept with analytical weaknesses, the idea of sustainability has provided a platform in which different stakeholders in tourism can interact, negotiate and reflect the consequences of their actions on the environment.

The major academic concern over the negative effects of tourism dates back to the research into tourism and recreation carrying capacity in the 1960s and 1970s (Saarinen, 2006). For almost two decades, the idea of carrying capacity formed the basis for approaching and managing negative impacts. However, after the period of enthusiasm, it was realised that carrying capacity could be problematic both in theory and in practice (O'Reilly, 1986; Wall, 1982). By the early 1990s, this issue was largely replaced in research and development discourses by the idea of sustainable tourism. However, surprisingly many challenges outlined for the present idea of sustainable tourism appear rather similar to past issues concerning carrying capacity. Therefore, it is easy to agree with Butler (1999: 15) when he asks critically whether the current ideas and discussions on the sustainable tourism are anything new. What seems to be new is the scale: unlike the rhetoric of sustainable tourism, the concept of carrying capacity does not imply global or intra- and inter-generational solutions but aims to offer more time- and space-specific answers at the local level. Therefore, it is occasionally interpreted as an application of sustainable tourism (Butler, 1996, 1999: 9): sustainable tourism involves the recognition of negative impacts and the need to manage them and carrying capacity has been one of the central frameworks within which such issues have been considered on a local scale (Lindberg *et al.*, 1997: 461).

Presently, the basic principles of sustainable development have been applied to tourism in many policy-level documents and programmes. For example, the United Nations World Tourism Organisation (UNWTO) (prior to 2003 as WTO) has made several official statements and declarations related to sustainable tourism (Table 1.2). According to the UNWTO (2004), sustainable tourism is a form of tourism that follows the principles

Table 1.2 UNWTO's (before 2003 as WTO) official statements and declarations related to sustainable development of tourism

Manila Declaration on World Tourism, 1980:   http://www.world-tourism.org/sustainable/doc/1980%20Manila-eng.pdf
Acapulco Documents on the Rights to Holidays, 1982:   http://www.world-tourism.org/sustainable/doc/1982%20ACAPULCO.PDF
Tourism Bill of Rights and Tourism Code, Sofia, 1985:   http://www.world-tourism.org/sustainable/doc/1985%20TOURISM%20 BILL%20OF%20RIGHTS.pdf
The Hague Declaration on Tourism, 1989:   http://www.world-tourism.org/sustainable/doc/THE%20HAGUE%20 DECLARATION.89.PDF
Lanzarote Charter for Sustainable Tourism, 1995 (jointly with UNEP, UNESCO, EU):   http://www.world-tourism.org/sustainable/doc/Lanz-en.pdf
Statement on the Prevention of Organized Sex Tourism, Cairo, 1995:   Agenda 21 for Tourism & Travel Industry, 1996:   http://www.world-tourism.org/sustainable/publications.htm#a21
Global Codes of Ethics for Tourism, 1999:   http://www.world-tourism.org/code_ethics/eng.html
Hainan Declaration – Sustainable Tourism in the Islands of the Asia-Pacific Regions (2002):   http://www.world-tourism.org/sustainable/doc/Hainan%20 Declaration-Dec%202000.pdf
Québec Declaration on Ecotourism, 2002:   http://www.world-tourism.org/sustainable/IYE/quebec_declaration/ eng.pdf
Djerba Declaration on Tourism and Climate Change, 2003:   http://www.world-tourism.org/sustainable/climate/decdjerba-eng.pdf

Source: UNWTO (2004)

of sustainability (see also WTO, 1993). The agency defines the concept broadly as tourism that is ecologically sound, economically viable and socially acceptable to the local communities in the long term; as also indicated by Clarke (1997), the UNWTO notes that sustainable tourism development guidelines and management practices are applicable to all forms of tourism and types of destinations, including CMT products and regions. Based on these premises, the specific goals of sustainable tourism should be (UNWTO, 2004) the following:

(1) Optimally use environmental resources that form a key element in tourism, maintaining essential ecological processes and helping to conserve nature and biodiversity.

(2) Respect the socio-cultural authenticity of local communities, conserve their cultural heritage and traditional values, and contribute to inter-cultural understanding.

(3) Ensure viable, long-term economic processes, providing socio-economic benefits to all stakeholders that are fairly distributed, including employment and income-earning opportunities and social services to local communities, and contributing to poverty alleviation.

Sustainable tourism development is seen as a continuous process that needs monitoring, prevention, management and governance of impacts. In addition, sustainable tourism should also maintain a high level of tourist satisfaction, and increase their awareness about sustainable development and related issues, and promote sustainable practices in tourist activities (UNWTO, 2004).

In spite of the policy-level attempts and perhaps as a result of conceptual problems, disagreements and the multi-dimensionality of both the concepts of tourism and sustainability (Butler, 1991; Lélé, 1991), many academic commentators have stated that no exact definitions of sustainable tourism exist (Hunter, 1995; Saarinen, 2006; Sharpley, 2000). Consequently, the notion has sometimes been understood as an ideology and point of view rather than an exact operational definition (Clarke, 1997). However, as contested as the concept may be, sustainability has become one of the key ideas and grounds for tourism studies and development. This starting point forms a challenging and interesting basis for the present volume, aiming at placing the international discussions of sustainability and tourism-environment into contexts of tourism development debates in southern Africa.

Sustainable Tourism Issues in Southern Africa

During the past decade, tourism and tourists have increasingly become a characteristic feature of contemporary southern African societies. The economic significance of tourism and the fact that tourism is developing fast mean that new attractions and facilities are constantly evolving to new areas. Places, regions and resources are being planned and transformed in order to attract more tourists, and also to attract investors in tourism. Even entire national economies in the region can be highly dependent on the needs of modern tourists and the tourist trade; in South Africa, the tourism industry has replaced the mining industry as the most important sector of economy. In Botswana tourism takes the second place (after the diamond industry), and in Namibia it is in the third position in economic rankings, for example.

Although the role of southern Africa and Africa in general has been relatively small in global tourism scene (Table 1.3), tourism development has been very positive from 1990s onwards, and southern Africa as a region has increasingly become an important destination for international tourists (see Rogerson, Chapter 2, this volume). According to UNWTO (2008), South Africa alone received about one-third of the sub-Saharan and one-fifth of Africa's international tourist arrivals in 2007 (excluding Egypt and Libya, see Rogerson, 2007: 363). In spite of the improved overseas accessibility, the majority of southern African tourism is regionally based and dominated by business and Visiting Friends and Relatives (VFR)-oriented visits. Rogerson presents a comprehensive overview of southern African tourism numbers in Chapter 2.

Along the increasing and developmental role of tourism industry in southern Africa, tourism has also emerged as an expanding field of academic study in universities and other institutions of tertiary education.

Table 1.3 International tourist arrivals (in millions) and the relative share (%) of arriving tourists by UNWTO regions. UNWTO's sub-regions do not include southern Africa

	1980	*1985*	*1990*	*1995*	*2000*	*2005*	*2007*
Africa	7.3	9.7	15.0	20.2	27.9	36.8	44.4
(%)	(2.6)	(3.0)	(3.3)	(3.6)	(4.1)	(4.6)	(4.9)
America	61.4	64.3	92.8	108.9	128.2	133.6	142.5
(%)	(21.5)	(19.7)	(20.3)	(19.3)	(18.8)	(16.5)	(15.8)
Asia and Pacific	21.5	31.3	54.6	81.4	109.3	156.7	184.3
(%)	(8.3)	(10.3)	(12.6)	(15.1)	(16.0)	(19.4)	(20.4)
Europe	186.0	212.0	282.7	338.4	393.5	441.6	484.4
(%)	(65.1)	(64.8)	(61.8)	(59.8)	(57.6)	(54.7)	(53.6)
Middle East	7.5	7.5	9.0	12.4	24.4	39.7	47.6
(%)	(2.6)	(2.3)	(2.0)	(2.2)	(3.6)	(4.9)	(5.3)
Total	285.9	327.1	457.2	565.4	683.0	808.0	903.0
(%)	(100)	(100)	(100)	(100)	(100)	(100)	(100)

Source: WTO (1999) and UNWTO (2008)

Tourism education is growing in the region, but according to Rogerson and Visser (2004) it is still in a relatively early phase (see Rogerson & Visser, 2007a). From the sustainable developmental perspective, for example, there is no comprehensive account of sustainable tourism that is easily accessible to higher education students or scholars who can serve to relate the southern African tourism context to the international tourism literature.

This book aims to draw connections between the core ideas of tourism and sustainability in southern African context. Naturally, it is not an easy task to integrate the terms of tourism and sustainability, both of which represent relatively complicated concepts with contested meanings, with a full review of a large and diverse region such as southern Africa. The main focus of discussions is set on the issues of tourism and sustainability in natural resource-dependent communities facing the processes of socio-economic and spatial transitions. These transition processes are long, gradual and are directly or indirectly linked to evolving tourism in localities. While recognising the key role of tourism in such processes, the social, cultural, economic and political contexts of tourism and local communities are also highlighted. Indeed, tourism does not happen in a vacuum and often the socio-spatial context makes the difference for the outcomes of tourism development.

In addition to the concept of sustainability, the term 'southern Africa' as a region is a contested idea that can be projected in numerous ways. By using the term 'southern Africa', we mainly refer to the continental and 'southern' countries of Botswana, Lesotho, Mozambique, Namibia, South Africa, Swaziland and Zimbabwe (Table 1.4). Although this practical selection covers all the UN geographical Southern Africa sub-region countries (with the additional countries of Mozambique and Zimbabwe), we recognise many of the limitations in our selection of cases and places (Figure 1.1), for example, many SADC countries such as Angola, Democratic Republic of the Congo, Malawi, Tanzania, Zambia and island states are mainly excluded from the discussion (however, see Rogerson, Chapter 2 in this book).

Although the stated focus – emphasised by the book's sub-title 'Local Communities and Natural Resources in Transition' – and issues discussed cover important aspects of tourism in southern Africa, evident and growing urban tourism and its connections to local communities and sustainability would have also been interesting to analyse. For example, the current issues related to tourism in townships or second-home tourism development would have provided fruitful grounds to link the connections between (urban) tourism, localities and sustainability (see Rogerson & Visser, 2007b). However, in order to cover the wider field including various themes and regions with illustrated cases another volume would clearly be required.

The purpose of this book is to explore the issues of sustainability in tourism, contributing especially to southern African, and also to

Table 1.4 Some of the key characteristics of the selected southern African countries

	Population (×1000) (2007)	Surface area (km²)	Protected areas[a] (% from surface area)	GDP (million USD) (2007)	GDP/capita (2007)	HIV[b] (%) (2007)	Adult literacy rate (%) (year)	Urban population (%) (2007)	Life expectancy[c]
Botswana	1882	581,730	30.2	9894	5259	23.9	81.2 (2003)	59	33.9
Lesotho	2008	30,355	0.2	1548	771	23.2	82.2 (2001)	19	34.3
Mozambique	21,397	801,590	8.6	8427	394	12.5	38.7 (1997)	36	41.8
Namibia	2074	824,292	14.6	6826	3291	15.3	85.0 (2001)	36	45.9
South Africa	48,577	1,221,037	6.1	277,825	5719	18.1	82.4 (1996)	60	44.1
Swaziland	1141	17,364	3.5	2895	2536	26.1	79.6 (2002)	25	29.9
Zimbabwe	13,349	390,757	14.7	4		15.3	89.5 (2004)	37	37.3

[a]Protected area ratio refers to totally or partially protected areas of at least 1000 ha, not including sites protected under local or provincial law.
[b]HIV (%): Adults living with HIV among 15–49 years old in 2007.
[c]Life expectancy: At birth 2005–2010.
Sources: UNCTAD (2008), UNEP (2008), UNMDG (2008), UN Statistics Division (2008) and UNWHO (2008)

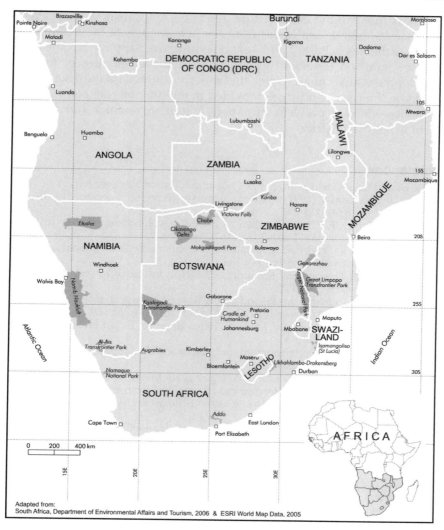

Figure 1.1   Location map of southern African countries and selected case study places and regions

acknowledge international tourism discussions on the role of natural resources and local communities in the context of tourism development. The book begins with an overview on tourism and sustainability issues in southern Africa, and offers the traditions and conceptual ground of sustainability in tourism development and studies. In addition, this contextual Part 1 introduces the developmental, planning and policy issues guiding and affecting tourism patterns in the region. Part 2 focuses on the role of nature and the use of natural resources in tourism, whereas Part 3 approaches the relating issues from the perspectives of local communities. The themes in Parts 2 and 3 are obviously overlapping, but the latter focuses more on the social issues, politics and economies of tourism development. In its final chapter, the book concludes with the discussions, based on the issues and cases provided in the previous chapters with an emphasis on future problem formations in tourism research and development in southern Africa.

References

Beck, U. and Beck-Gernsheim, E. (2001) *Individualization*. London: Sage.

Berry, S. and Ladkin, A. (1997) Sustainable tourism: A regional perspective. *Tourism Management* 18 (4), 433–440.

Binns, T. and Nel, E. (2002) Tourism as a local development strategy in South Africa. *The Geographical Journal* 168 (3), 235–247.

Bramwell, B. and Lane, B. (1993) Sustaining tourism: An evolving global approach. *Journal of Sustainable Tourism* 1 (1), 1–5.

Britton, S. (1982) The political economy of tourism in the Third world. *Annals of Tourism Research* 9, 331–358.

Britton, S. (1991) Tourism, capital and place: Towards a critical geography of tourism. *Environment and Planning D: Society and Place* 9 (4), 451–478.

Butler, R.W. (1990) Alternative tourism: Pious hope or Trojan Horse? *Journal of Travel Research* 28 (3), 40–45.

Butler, R. (1991) Tourism, environment, and sustainable development. *Environmental Conservation* 18 (3), 201–209.

Butler, R. (1996) The concept of carrying capacity for tourism destinations: Dead or merely buried? *Progress in Tourism and Hospitality Research* 2 (3–4), 283–293.

Butler, R. (1999) Sustainable tourism: A state-of-the-art review. *Tourism Geographies* 1 (1), 7–25.

Cater, E. (1993) Ecotourism in the third world: Problems for sustainable development. *Tourism Management* 14 (1), 85–90.

Clarke, J. (1997) A framework of approaches to sustainable tourism. *Journal of Sustainable Tourism* 5 (2), 224–233.

Duffy, R. (2002) *A Trip Too Far: Ecotourism, Politics and Exploitation*. London: Earthscan.

Elliott, J.A. (1994) *An Introduction to Sustainable Development*. London: Routledge.

Fennell, D. (1999) *Ecotourism: An Introduction*. London: Routledge.

Goudie, S.C., Khan, F. and Kilian, D. (1999) Transforming tourism: Black empowerment, heritage and identity beyond apartheid. *South African Geographical Journal* 81 (1), 21–33.

Government of South Africa (1996) White paper: Development and promotion of tourism in South Africa. Department of Environmental Affairs and Tourism, June 1996, 72pp.

Gössling, S. and Hall, C.M. (2005) An introduction to tourism and global environmental change. In S. Gössling and C.M. Hall (eds) *Tourism and Global Environmental Change* (pp. 1–33). London: Routledge.

Hall, C.M. (1998) Historical antecedents of sustainable development and ecotourism: New labels on old bottles? In C.M. Hall and A.A. Lew (eds) *Sustainable Tourism: A Geographical Perspective* (pp. 13–24). New York: Longman.

Hall, C.M. (2000) *Tourism Planning: Policies, Processes and Relationships*. Essex: Pearson Education Limited.

Hall, C.M. (2005) *Tourism: Rethinking the Social Science of Mobility*. Harlow: Prentice-Hall.

Hall, C.M. and Lew, A.A. (1998) The geography of sustainable tourism: Lessons and prospects. In C.M. Hall and A.A. Lew (eds) *Sustainable Tourism: A Geographical Perspective* (pp. 199–203). New York: Longman.

Holden, A. (2003) In need of new environmental ethics for tourism. *Annals of Tourism Research* 30 (1), 94–108.

Holden, A. (2007) *Environment and Tourism*. London: Routledge.

Hughes, G. (1995) The cultural construction of sustainable tourism. *Tourism Management* 16 (1), 49–59.

Hunter, C.J. (1995) On the need to re-conceptualise sustainable tourism development. *Journal of Sustainable Tourism* 3 (3), 155–165.

Inskeep, E. (1991) *Tourism Planning: An Integrated and Sustainable Development Approach*. New York: Van Nostrand Reinhold.

Lélé, S. (1991) Sustainable development: A critical review. *World Development* 19 (6), 607–621.

Lindberg, K., McCool, S. and Stankey, G. (1997) Rethinking carrying capacity. *Annals of Tourism Research* 24 (2), 461–465.

Liu, Z. (2003) Sustainable tourism development: A critique. *Journal of Sustainable Tourism* 11 (4), 459–475.

Macbeth, J. (2005) Towards an ethics platform for tourism. *Annals of Tourism Research* 32 (4), 962–984.

Mathieson, A. and Wall, G. (1982) *Tourism: Economic, Physical and Social Impacts*. New York: Longman.

Meadows, D.H., Meadows, D., Randers, J. and Behrens, W.W. III (1972) *The Limits to Growth: A Report for the Club Rome's Project on Predicament of Mankind*. London: Earth Island Limited.

Milne, S. (1998) Tourism and sustainable development: The local–global nexus. In C.M. Hall and A.A. Lew (eds) *Sustainable Tourism: A Geographical Perspective* (pp. 35–48). New York: Longman.

Ministry of Environment and Tourism (2005) *A National Tourism Policy for Namibia*. Windhoek: Ministry of Environment and Tourism.

Mowforth, M. and Munt, I. (1998) *Tourism and Sustainability: A New Tourism in the Third World*. London: Routledge.

O'Reilly, A.M. (1986) Tourism carrying capacity: Concepts and issues. *Tourism Management* 7 (4), 254–258.

Pigram, J.J. (1990) Sustainable tourism – policy considerations. *The Journal of Tourism Studies* 1 (2), 2–9.

Poon, A. (1993) *Tourism, Technology and Competitive Strategies*. Wallingford: CAB International.

Redcliffe, M. (1987) *Sustainable Development: Exploring the Contradictions*. London: Methuen.

Redcliffe, M. and Woodgate, G. (1997) Sustainability and social construction. In M. Redcliffe and G. Woodgate (eds) *The International Handbook of Environmental Sociology* (pp. 55–67). Cheltenham: Edward Elgar.

Rogerson, C. (1997) Local economic development and post-apartheid reconstruction: The case of South African cities. *Singapore Journal of Tropical Geography* 18 (2), 175–195.

Rogerson, C. (2007) Reviewing Africa in the global tourism economy. *Development Southern Africa* 24 (3), 361–380.

Rogerson, C. and Visser, G. (2004) Tourism and development in post-apartheid South Africa: A ten year review. In C.M. Rogerson and G. Visser (eds) *Tourism and Development Issues in Contemporary South Africa*. Pretoria: Africa Institute of South Africa.

Rogerson, C. and Visser, G. (2007a) Tourism research and urban Africa: The South African experience. In C. Rogerson and G. Visser (eds) *Urban Tourism in the Developing World: The South African Experience*. New Brunswick: Transaction Publishers.

Rogerson, C. and Visser, G. (eds) (2007b) *Urban Tourism in the Developing World: The South African Experience*. New Brunswick: Transaction Publishers.

Saarinen, J. (2006) Traditions of sustainability in tourism studies. *Annals of Tourism Research* 33 (4), 1121–1140.

Saarinen, J. (2007) Cultural tourism, local communities and representations of authenticity: The case of Lesedi and Swazi cultural villages in Southern Africa. In B. Wisheitemi, A. Spenley and H. Wels (eds) *Culture and Community: Tourism Studies in Eastern and Southern Africa* (pp. 140–154). Amsterdam: Rozenberg.

Scheyvens, R. (2002) *Tourism for Development: Empowering Communities*. Harlow: Prentice Hall.

Sharpley, R. (2000) Tourism and sustainable development: Exploring the theoretical divide. *Journal of Sustainable Tourism* 8 (1), 1–19.

Spangenberg, J.H. (2005) Will the information society be sustainable? Towards criteria and indicators for sustainable knowledge society. *International Journal of Innovation and Sustainable Development* 1 (1–2), 85–102.

Telfer, D. and Sharpley, R. (2007) *Tourism and Development in the Developing World*. London and New York: Routledge.

Turner, L. and Ash, J. (1975) *The Golden Hordes: International Tourism and the Pleasure Periphery*. London: Constable.

Urry, J. (1990) *The Tourist Gaze: Leisure and Travel in Contemporary Societies*. London: Sage.

Wall, G. (1982) Cycles and capacity: Incipient theory of conceptual contradiction? *Tourism Management* 3 (3), 188–192.

Wall, G. (1997) Sustainable tourism – unsustainable development. In S. Wahab and J.J. Pigram (eds) *Tourism, Development and Growth: The Challenge of Sustainability* (pp. 33–49). London: Routledge.

WCED (1987) *Our Common Future*. Oxford: Oxford University Press.

van Veuren, E.J. (2001) Transforming cultural villages in the spatial development initiatives of South Africa. *South African Geographical Journal* 83 (2), 137–148.

Williams, A.M. and Shaw, G. (1998) Tourism and the environment: Sustainability and economic restructuring. In C.M. Hall and A.A. Lew (eds) *Sustainable Tourism: A Geographical Perspective* (pp. 49–59). New York: Longman.

UNCTAD (2008) *Handbook of Statistics*. On WWW at http://stats.unctad.org/handbook/ReportFolders/ReportFolders.aspx?CS_referer=&CS_ChosenLang=en. Accessed 1.12.2008.

UNMDG (2008) *Millennium Development Goals*. On WWW at http://www.mdgmonitor.org/map.cfm?goal=4&indicator=0&cd=. Accessed 1.12.2008.

UNEP (2008) *Globalis*. On WWW at http://globalis.gvu.unu.edu/. Accessed 1.12.2008.

UN Statistics Division (2008) *Environment Statistics – Country Snapshots*. On WWW at http://unstats.un.org/unsd/environment/Questionnaires/country_snapshots.htm. Accessed 1.12.2008.

UNWHO (2008) *Adult Literacy Rate*. On WWW at http://www.who.int/whosis/en/index.html. Accessed 1.12.2008.

UNWTO (2004) *Sustainable Development of Tourism Conceptual Definition*. On WWW at http://www.world-tourism.org/frameset/frame_sustainable.html. Accessed 10.2.2006.

UNWTO (2008) *UNWTO World Tourism Barometer* 6 (2). On WWW at http://unwto.org/facts/eng/pdf/barometer/UNWTO_Barom08_2_en_LR.pdf. Accessed 10.10.2008.

World Travel and Tourism Council (2007) *Botswana: The Impacts of Travel and Tourism on Jobs and the Economy*. London: WTTC.

WTO (1993) *Sustainable Tourism Development: Guide for Local Planners*. Madrid: Word Tourism Organization.

WTO (1999) *Tourism 2020 Vision, Volume 7, Global Forecast*. Madrid: Word Tourism Organization.

Chapter 2

Tourism Development in Southern Africa: Patterns, Issues and Constraints

CHRISTIAN M. ROGERSON

Introduction

The task in this chapter is to furnish an introduction to the key issues surrounding tourism development across the 'southern African region'. The United Nations World Tourism Organization (UNWTO) defines southern Africa as comprising the five countries of Botswana, Lesotho, Namibia, South Africa and Swaziland. The region of southern Africa is defined here more broadly than the UNWTO classification to include four additional countries, namely Malawi, Mozambique, Zambia and Zimbabwe, which in UNWTO classifications fall into the eastern African region. All nine countries share in common the membership of the Southern African Development Community as well as the 14-country Regional Tourism Organization of Southern Africa (RETOSA), which has the mandate to market and promote the region in terms of forging a concrete destination identity in the international tourism market.

This analysis unfolds through four uneven sections of material. First, tourism in southern Africa is situated within a wider African and global context. Second, an overview of the key tourism products and branding of the region is provided. Third, an analysis is undertaken of the structure of the regional tourism economy and of spatial patterns of tourism flows. Finally, the last section highlights certain constraints upon tourism development in the region that affect the contribution of tourism towards sustainability in the region.

Southern Africa: The Global and African Context

According to UNWTO data, during 2005 international tourism arrivals for the first time were in excess of 800 million. Africa's share of global tourism was recorded as 5.6%, a share that has been described as 'a drop in the ocean' (Mitchell & Ashley, 2006: 1). Nevertheless, many observers assert Africa's tourism potential to be significant, albeit underutilized or underdeveloped (Christie & Crompton, 2001; Elliott & Mann, 2006; Naude & Saayman, 2004). For example, Mitchell and Ashley (2006: 4) affirm that 'international tourism is already important for Africa – and is likely to grow' in future. Several writers draw attention to the fact that the tourism sector can contribute towards sustainability as it can represent a solution for both economic growth and poverty alleviation in Africa (Ashley & Mitchell, 2006; Elliott & Mann, 2006; Visser, 2006; World Bank, 2006). Others highlight that tourism in the developing world can contribute more broadly to achievement of the goals of the United Nations Millennium Project (Telfer & Sharpley, 2008).

Behind the expansion of African tourism and of the importance of tourism for sustainability in Africa, the World Bank (2006) identifies several factors. It argues that

> Africa offers abundant diversity; access from the major source markets (Europe, USA and Asia) to African destinations is improving through hubs like Johannesburg, Nairobi, Addis Ababa, and Lagos; there is a major consumer shift away from pre-paid packages to independently organized travel, which increases the distribution options for African products; and Africa itself is a growing market that is driving demand for domestic and regional tourism, both for leisure and business. (World Bank, 2006: 1–2)

This optimism appears justified because during 2005 Africa achieved the best performance in terms of growth of international arrivals of any of the UNWTO regions (UNWTO, 2006). Moreover, since 1990 Africa has been steadily increasing its share of the highly competitive global tourism economy with Europe, the major source region for long-haul traffic (Mitchell & Ashley, 2006). That said, it must be understood that intra-African travel or 'regional tourism' is by far the largest single sub-sector within the category of 'international tourism arrivals' in Africa (Rogerson, 2007a). Overall, Mitchell and Ashley (2006: 3) record that 'despite growth in global tourism receipts over the past decade of about 3 percent each year tourism in Africa has increased its global market share of international arrivals'. Future prospects for tourism in Africa are viewed as promising: 'tourism in Africa should continue to grow both absolutely and in terms of global market share' (Mitchell & Ashley, 2006: 3).

Currently, the geographical patterns of tourism across Africa are, however, markedly uneven. At the level of individual countries, tourism in Africa is strongly concentrated upon the continent's 'big four' destinations: Egypt, South Africa, Tunisia and Morocco (Naude & Saayman, 2004;

Rogerson, 2007a). At the sub-regional level, North Africa is seen as the leading destination followed closely by the UNWTO southern Africa region. Together, the nine nations of the southern African region under scrutiny in this volume account for a total of 13.5 million international arrivals (2005) or 29.6% of all Africa's international tourism arrivals.

Key Tourism Products and Promotion

The major tourism products and destinations of southern Africa are shown in Figure 2.1. According to RETOSA (2008), tourism in southern Africa represents both 'the best of Africa' and 'the essence of Africa'. It proclaims that the region offers a diverse range of travel destinations with 'everything from the definitive safari to exquisite hospitality, adrenalin-pumping adventure and unique cuisine' (RETOSA, 2008).

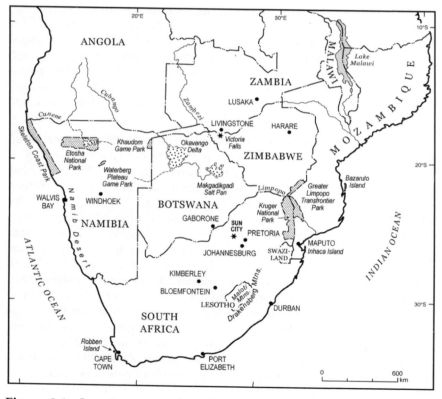

Figure 2.1 Location map of southern African region, including major tourism attractions

In terms of the nine focus countries, the promotional emphasis is firmly upon 'the coral coastlines of Mocambique', 'the vineyards, bustling cities and big game reserves of South Africa', 'Botswana's Okavango Delta – the jewel of Africa', 'the grand mountains of Lesotho', 'the unparalleled Victoria Falls shared by Zambia and Zimbabwe' (Box 2.1) and the 'natural wonders' of Lake Malawi and the Namib Desert. Significantly, perhaps, RETOSA is silent on Swaziland's attractions for international tourism. The website of the Swaziland Tourism Authority is also somewhat low key in its marketing, stating only that 'If you travel through the country you will discover, as others have, that Swaziland has a quaint offering of diverse vegetation at every turn, which one would probably expect from a larger country' (Swaziland Tourism Authority, 2008).

The nature of RETOSA's promotional efforts confirms that in terms of the international tourism market, the southern African region is branded as a destination primarily for 'alternative' rather than mass-packaged tourism. Four core reasons to visit the region are identified. These are the diversity of attractions, adventure ('Southern Africa offers the best adventure one can get'), affordability in terms of local currencies versus the European currencies or the US$, and, most importantly, the region's 'big 5' game attractions (RETOSA, 2008). The activities that are highlighted for visitors are safaris, river adventures, hunting, shark diving, scuba diving, bungee jumping, whale watching, birding, golfing, touring historical sites and overland exploration. Emphasis is strongly on the strength of the region's tourist products for ecotourism, adventure tourism and cultural/rural tourism, which, it has been observed, are some of the fastest expanding segments of global tourism (Telfer & Sharpley, 2008; Visser, 2006).

Box 2.1   Tourism in the Victoria Falls: A Zimbabwean Perspective

HARETSEBE MANWA

Victoria Falls is the world's largest waterfall. It is located in the northwestern part of Zimbabwe on the border with Zambia. According to Masocha (2006), the Victoria Falls resort handles about 30,000 tourists a month. The town is very important to the economy of the country in terms of bringing in foreign currency. As shown in Table 2.1, the resort is completely dependent on international tourists. The aim of this case study box is to discuss the sustainability of tourism at the resort taking into consideration both environmental and socio-economic impacts of tourism on the town.

Table 2.1 Occupancies by region 2005 compared with 2004

Region	Bed occupancy rates			Room occupancy rates			% Client composition	
	Total beds	2005 (%)	2004 (%)	Total rooms	2005 (%)	2004 (%)	Local	Foreign
	11,282			5657				
Bulawayo	1278	37	34	586	50	51	83	7
Harare	3945	29	27	2146	41	39	95	5
Hwange	337	12	10	168	16	13	90	10
Kariba	849	23	23	406	32	35	93	7
Masvingo	339	26	37	255	38	56	95	5
Midlands	505	28	31	227	40	44	99	1
Mutare/ Vumba	787	29	33	399	42	47	87	13
Nyanga	563	22	30	272	33	43	97	3
Beitbridge	441	49	55	212	67	70	92	8
Victoria Falls	2238	26	23	986	31	29	39	71
Average		28	30		39	43	87	13

Source: ZTA (2005)

Motivation to Visit Victoria Falls

Being a World Heritage site is one of the major motivators to visit Victoria Falls, which is one of the Seven Natural Wonders of the World (Manwa, 2007). Victoria Falls is also the icon of Zimbabwe, which has become a brand in international tourism market. In general, motivation to visit the resort can be classified as leisure and adventure using the activities undertaken by tourists. The leisure tourists visit Victoria Falls to rest and relax in a friendly atmosphere. They engage in the following activities: boat cruising to watch the sunrise or sunset and viewing the river-based wildlife such as hippopotamus, crocodiles, fish and birds found at the banks of the Zambezi River. These tourists are nature lovers whose main motivation to visit the resort is to admire nature (Boyd & Butler, 1996), including the rain forest, the falls themselves or go on game drives and photographic safaris. The town of Victoria Falls itself has plenty of wildlife, Kudu, elephants, giraffes and other smaller animals.

The second category of tourists to the resort is the adventure tourists. These are tourists who want to engage in activities that raise the adrenaline. The most popular activities are white water rafting and bungee jumping. Because of the variety of activities a tourists can engage in, it has the highest density of tour operators in Zimbabwe trying to catch on the tourists flocking to the town.

Environmental Impacts

Hall and Page (1999) have argued that tourism is dependent on the environmental quality and that its success or failure is closely associated with an attractive, healthy and pleasant natural environment. If the landscape is somehow negatively affected, tourists will stay away, even when damage is only perceived and not real. Many tourists believe that an environment is healthy as long as it looks 'clean and green'. It is therefore clear that the protection of the natural and cultural resources upon which tourism is based is very important for the sustainability of a resort (Hall & Page, 1999: 152).

Budowski (1976) on the other hand came up with three relationships that exist between tourism and the environment. These relationships are conflict, coexistence and symbiosis. The conflict relationship shows that the view of maintaining a good-quality environment contradicts with the view of expanding tourism industry. The expansion of tourism industry will cause adverse effects on the environment and therefore result in the reduction of the quality of the environment since it involves activities that can harm the environment, for example, construction of general infrastructure such as roads and airports, and tourism facilities including resorts, hotels, restaurant, shops and golf courses (UNEP, 2002).

However, the coexistence relationship shows that tourism and the environment exist together at the same time; when there is good environment, landscape, cultural and natural resources, the tourism industry will survive. The symbiosis relationship shows that tourism industry is dependent on the environment for its survival; on the other hand, the environment depends on tourism. This is achieved through raising awareness of environmental value (UNEP, 2002).

As has been eloquently espoused by Butler (1990), tourism is an exploitative industry whereby operators are always ready to move into an area as long as it is attractive and to quickly move out once it has lost its pristine environmental appeal. This point is confirmed by Thomas *et al.* (2005) who agree that once destinations have reached their carrying capacity (Butler, 1980) it can result in the erosion of the environment (UNEP, 2002).

Sanyanga and Masundire (1999) studies on Victoria Falls show that sewage effluent is discharged into the Zambezi River via the tributary, the Masure River. There is poor effluent quality in Victoria Falls, which affects downstream users including water-based tourism. Their study found faecal contamination at 10 km downstream from Victoria Falls outflow. They also observed that a few boats were dumping raw sewage into the Zambezi River and that there was lack of enforcement and clear regulations regarding sewage disposal.

Other environmental impacts of tourism on Victoria Falls, especially along the Zambezi River, are changes in the behaviour of some animals and vulnerable bird species. The African Skimmer and Pels fishing owl birds have been adversely affected by usage of powerboats. Their feeding and nesting sights have been adversely affected by disturbances from boat traffic and human pressure. This has resulted in the behavioural change and loss of such species (Hoole, 2001). Some operators also toss waste food into the river, which attracted crocodiles and fish. Crocodiles pose danger to people as they tend to attack them at cruise boat jetties in search of food (Hoole, 2001). Continuous usage of the Zambezi River has also changed the behaviour of wildlife such as elephants, buffalo and hippopotamus, which have become aggressive because of continuous disturbance by boat cruises.

Carrying Capacity

It is also believed that the river has reached its maximum carrying capacity. As a result, the river is characterized by crowding, which tends to obstruct tourists' view who are there to enjoy either the Zambezi River sunset or sunrise. Heath (1986) observed more than 20 years ago that the resort had reached its carrying capacity in terms of infrastructure development. Despite this realization, the resort saw massive infrastructure developments whereby international hotels brands built accommodation facilities. This has the obtrusive effect that it interferes with the scenic beauty of the river and also with the movement of wildlife to the Zambezi River. This will negatively impact on visitors' satisfaction. Noise pollution is another problem at the resort, for example low-flying aircraft and the helium balloon rides over the falls.

Enclave Tourism

The resort, like most other tourists destinations in developing countries, was until recently owned and controlled by multinational corporations or local corporations and had no impact on the livelihoods of local people. This became a problem whereby criminal activities targeted at tourism were rife. Instead, local people and other local migrants resorted to hawking, selling of curios (Matose *et al.*, 1997) or dumping scavenging (Masocha, 2006). Another interesting development in the town was the use of the US dollar for trading. Instances of tourists being charged 20 US dollars for a meal were not uncommon.

Other elements of the resort that differentiated it from the rest of the country were the dress code and the language of communication. Young men and women living and working at the resort emulated the American and British tourists in the form of language and dress.

Conclusion

Victoria Falls like the rest of the country is experiencing a decrease in tourist arrivals and spending. The resort is seeing more day trippers, who either fly into the resort from destinations in South Africa or are self-driven and have not much impact on the local economy. For example, there were 1,43,685 visitors who did not spend a night in Zimbabwe. Forty-five percent arrived through Beitbridge border post, while 22% entered through Victoria Falls border post (ZTA, 2005). On a same day, international visitors were 557 from the USA, 506 from Malaysia and 551 from Italy. The country is no longer attractive to visitors. Perhaps this is a blessing in disguise to avert some of the environmental dangers that were visible in the 1990s. The decline of the destination would help it rejuvenate its natural resources. In an effort to rejuvenate the resort, the Victoria Falls Council together with other stakeholders would have to come up with management practices that will ensure sustainability of the resort, including banning some activities in specific fragile and sensitive areas. More importantly, environmental education should be implemented for all the stakeholders including international tourists.

The promotional efforts of national tourism organizations in the region likewise target alternative tourism. For example, Botswana is marketed internationally as a destination for low-volume, high-value safari visitors (Mbaiwa, 2003, 2004, 2005a) and Zambia is marketed as 'the real Africa' for its game parks and adventure tourism activities around Livingstone (Rogerson, 2003a). Tourism promotion in Lesotho involves linking to South African tourism, so that international visitors to South Africa would enter the country on a short visit, mainly for adventure trips, hiking in the mountains or pony trekking (Mashinini, 2003). Even for South Africa, the core tourism destination within southern Africa, the marketing emphasis is upon the attractions of landscape, wildlife and culture

(Rogerson & Visser, 2004). Critically, as Cornelissen (2005) argues, South Africa is not viewed or marketed as a destination for mass-packaged beach tourism. Recently, however, a number of other kinds of tourism have attracted policy focus, including events and sports tourism, the latter linked to the hosting of the 2010 FIFA World Cup.

The national Department of Trade and Industry (DTI) has sought to identify certain special interest or 'niche tourism' products which might be targeted in order to diversify the product mix and thus enhance further the appeal of South Africa for international tourism. Backpacking and youth travel, business tourism, birding and community-based route tourism are the initial foci of DTI's promotional efforts. It is argued that niche tourism in South Africa 'can contribute towards the tourism sector's objectives of increasing tourist's length of stay, spend, geographical distribution, volumes, reducing seasonality and driving transformation in the sector' (DTI, 2007: 6). In particular, backpacker tourism is considered by DTI to be a particularly attractive segment or niche market in which there is potential for future expansion both in South Africa and into surrounding countries of the region (Rogerson, 2007b, 2007c). Other growing forms of niche tourism in South Africa that so far have been overlooked by the DTI include gay tourism (Visser, 2002, 2003a, 2003b) and health tourism.

Beyond the alternative tourism focus for attracting long-haul international tourists, it must be recognized that within the region of southern Africa there are tourism products that are aimed largely at the *domestic or regional tourists* rather than the long-haul international tourists. In South Africa, casinos, exhibition halls, convention centres and beaches are some of the major tourism products that are largely for the domestic tourism market (Rogerson, 2003b; Rogerson & Visser, 2004, 2007). For the regional market, the importance of shopping malls must also be noted (Rogerson, 2004; Rogerson & Kiambo, 2007). With the exception of South Africa, domestic tourism across the region centres mainly upon business tourism rather than leisure tourism. In Botswana or Zambia, for example, an important segment of urban tourism economy of Gaborone or Lusaka relates to the activity of business tourism (Mbaiwa *et al.*, 2007).

On the basis of the economic growth that has been experienced in other southern African countries during the past decade, there is also emerging parallel groups of people who enjoy sufficient income and desire to travel both within their countries and to regional destinations (Mbaiwa *et al.*, 2007). The growth of regional travel within southern Africa is an important new phenomenon in the regional tourism landscape. Typically, Figure 2.2 illustrates the expansion of organized travel groups from Botswana to other destinations in the region.

South Africa has the most well-established and largest domestic tourism economy across the region (Rogerson & Lisa, 2005). In racial terms, the greatest expenditure component of domestic tourism is accounted for

Figure 2.2 The promotion of regional tourism from Botswana to South Africa and Mozambique

by white South Africans. Nevertheless, there is a rapidly growing black middle class, which is changing the nature of domestic tourism in post-apartheid South Africa. Further, since 2002 the national government has been energetically supporting the rise of domestic tourism through its 'Sho't Left' promotional campaign (Rogerson & Lisa, 2005; Rogerson & Visser, 2007). Domestic tourism is a significant stabilizing element in the South African tourism economy, albeit its contribution in value terms is much less than that of international tourism (Rogerson & Visser, 2004).

Tourism Growth and Patterns

Using data for 2005, Figures 2.3 and 2.4 show the relative importance of tourism in the nine countries of southern Africa in terms of both the volume of international tourism arrivals and international tourism receipts. The maps disclose clearly the massive dominance of South Africa within the regional tourism economy. South Africa's core position in the region is underscored by the fact that the country accounts for 55.6% of the region's international tourism arrivals and 83.7% of international tourism receipts.

Taken together, Tables 2.2 and 2.3 reveal a number of important trends that characterize the tourism economy of the southern African region.

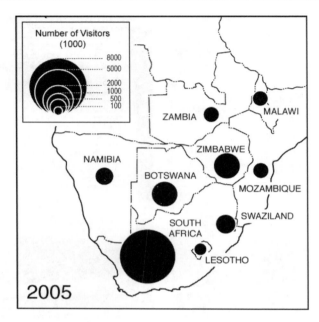

Figure 2.3 International tourism arrivals by country, 2005
Source: UNWTO

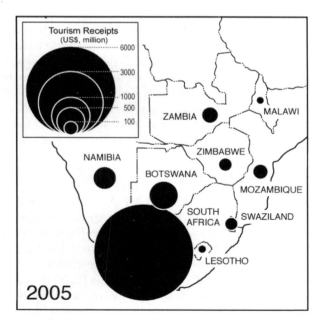

Figure 2.4 Receipts from international tourism
Source: UNWTO data

Table 2.2 International tourism arrivals by country 1990–2005 (1000s)

Country	1990	1995	2000	2005	*% Share of 2005 regional arrivals*
Botswana	543	521	1104	1523 (2004)	11.3
Lesotho	242	209	302	304	2.2
Malawi	130	192	228	471 (2004)	3.5
Mozambique	NA	NA	NA	470 (2004)	3.5
Namibia	NA	272	656	695 (2003)	5.1
South Africa	1029	4684	6001	7518	55.6
Swaziland	263	300	281	459 (2004)	3.4
Zambia	141	163	457	515 (2004)	3.8
Zimbabwe	636	1416	1967	1559	11.5

Source: UNWTO 2006 Annex 4
Note: For share of regional arrivals, most recent data for each country are used.

Since 1990, there has been a remarkable growth in the volume and significance of tourism across the region. In 1990, international tourism arrivals across the region were approximately 3 million; by 2005 the regional total was 13.2 million. In 1990, international tourism receipts were in the order of US$2200 million, whereas by 2005 this had escalated to US$8778 million.

Table 2.3 International tourism receipts by country 1990–2005 (US$ million)

Country	1990	1995	2000	2005	*% Share of regional tourism receipts*
Botswana	117	162	222	562	6.4
Lesotho	17	27	24	30	0.3
Malawi	16	17	25	26	0.3
Mozambique	NA	NA	74	130	1.5
Namibia	85	278	160	348	4.0
South Africa	1832	2125	2675	7327	83.7
Swaziland	30	48	37	95 (2004)	1.1
Zambia	41	47	111	161 (2004)	1.8
Zimbabwe	60	145	125	99	1.1

Source: UNWTO 2006 Annex 11
Note: For share of regional receipts, most recent data for each country are used.

Accordingly, in relation to both numbers of arrivals and receipts from international tourism, a fourfold increase is recorded during the period 1990–2005 for the region as a whole.

Several factors underpin this trend in growth. The most significant is the 1994 democratic transition in South Africa, which allowed the country's full re-entry into international tourism after many years of sanctions and boycotts (Rogerson & Visser, 2004). The 'Mandela Boom' in international tourism arrivals to South Africa is reflected in a remarkable 19.3% annual growth of tourism arrivals between 1990 and 2000. Other contributory factors in the region's overall growth performance have been the steady rates of expansion of tourism arrivals in Botswana, Zambia and, at least until 2000, in Zimbabwe. Of note, however, is that Zambia's popularity as a tourist destination has been expanding post-2000 due, in part, to the political problems in neighbouring Zimbabwe with both international and regional tourists electing to visit Zambia rather than Zimbabwe (Rogerson, 2003a).

Finally, in terms of tourism growth across the region, attention must be drawn to the emergence, during the 1990s, of Namibia (after independence) and (later) of Mozambique as new destinations for international tourism, the latter after many years of civil war and strife (Kiambo, 2005; FIAS and OECD Development Centre, 2006). Lesotho, Swaziland and Malawi occupy the bottom places of the regional league Table of southern Africa in terms of tourism development indicators. In the case of Malawi, however, a modest growth of international tourism arrivals has occurred post-2002 as a result of a marketing campaign, which was undertaken by the Ministry of Tourism, Parks and Wildlife to popularize the country in key European markets as an alternative tourism destination in Africa (Euromonitor International, 2007a).

Although rates of tourism growth have fluctuated across the region, there has been an upward trajectory in both arrivals and receipts in six of the eight countries throughout the period 1990–2005. Zimbabwe is the most exceptional case with the country experiencing a 'hollowing out' of the national tourism economy. Tourism has plummeted alongside the country's economic crisis and growing political turmoil; between 1995 and 2005 receipts from international tourism collapsed by nearly one-third (Rogerson, 2007a). Evidence for the period 2006–2007 suggests that the downward trend has been especially stark. Euromonitor International (2007b) records that for 2006–2007 the overall market witnessed a 39% decline in long-haul international arrivals with the sharpest declines from the traditional tourist markets of the United Kingdom, USA and Germany. In addition, tourist arrivals from Africa also exhibit a downward spiral, with an 11% decline (Euromonitor International, 2007b). In addition to Zimbabwe, the Swaziland tourism economy also shows signs of downturn, according to a recent analysis by Euromonitor International (2007c). Their analysis indicates strong signals of stagnation and decline in

Swaziland tourism. The downturn in the country's tourism economy has not been assisted by the blacklisting in March 2006 by the European Union of several Swazi airlines that were declared to be 'unsafe'.

As has been noted above, a striking aspect of the southern African tourism region is the post-1990 surge, which has taken place in regional travel (Rogerson, 2004; Rogerson & Kiambo, 2007). The international experience shows that the emergence and expansion of regional tourism is inseparable from the growth of domestic tourism in developing countries (Ghimire, 2001a, 2001b). The same sets of factors that gave rise to mass tourism in the countries of the First World are now providing the basis for what Gladstone (2005) styles as a 'new kind of tourism' in the developing world, including the southern Africa region (see Rogerson, 2004). The magnitude and significance of regional tourism is made clear by unpacking the geography of international tourism in South Africa. Overall, of the total of 7.37 million foreign arrivals in South Africa for 2005, continental Africa is the source of 5.36 million or 72.7% (Rogerson & Visser, 2006).

The country sources of international tourism arrivals for South Africa in 2005 are mapped in Figure 2.5. This indicates the enormous role played by African tourism in the country's tourism economy. It is evident that Africa, South Africa's natural hinterland, provides the baseload or backbone of the country's international tourism economy (Rogerson & Kiambo, 2007; Rogerson & Visser, 2006; Saayman & Saayman, 2003). Together, the leading five individual source countries – Lesotho, Swaziland, Botswana,

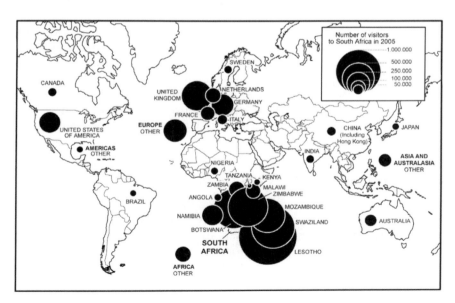

Figure 2.5 Source of international tourism arrivals in South Africa, 2005

Zimbabwe and Mozambique – account for 4.73 million arrivals or 64.2% of all 2005 international tourism arrivals into South Africa. The analysis of Euromonitor International (2007d) reinforces the point that behind South Africa's impressive rise in international tourism arrivals is the continued expansion of arrivals from neighbouring states in the region.

As pointed out by Rogerson and Visser (2006: 203), 'South Africa's two groups of 'international' tourists visit the country for very different reasons'. The majority of long-haul arrivals from Europe or North America are visiting South Africa for purposes of holiday or leisure. In contrast, only 20% of all African arrivals in South Africa in 2005 were travelling for leisure purposes. According to South African Tourism data, 45.6% of all African visitors were travelling to South Africa for business purposes and 24.7% is recorded as travelling to South Africa for Visiting Friends and Relatives (VFR) tourism (see Rogerson & Kiambo, 2007). For regional travel, travel over land is often a preferred mode of transport to air arrival, particularly among the groups of VFR tourists and cross-border shoppers into South Africa.

Of special importance in terms of the category of business travel to South Africa is the group of shoppers and cross-border traders from African destinations. It is calculated that one-third of all African tourists in South Africa are visiting the country for the main reason of shopping. The largest number of cross-border shoppers are drawn to South Africa from Mozambique, Zimbabwe, Botswana and Swaziland, and collectively represent the major share of the total African land market for South African tourism. The focal point for regional shoppers is the city of Johannesburg (Rogerson, 2002). The metropolitan authorities in Johannesburg are seeking now to harness the potential benefits of this market niche in shopping tourism for regenerating parts of the city, a development strategy which is earning it the popular acronym of 'Jobai' as the city seeks to carve out a parallel role in shopping tourism in Africa to that of Dubai in the Middle East (see Henderson, 2006). It must be understood, however, that the largest share of shopping tourists in Johannesburg are markedly different in spending power to those in Dubai, as they are mainly low-income cross-border traders or 'informal businesses'.

Constraints on Tourism Development

As is the case for much of the African continent, tourism in the southern African region is considered now to have much potential for contributing towards promoting economic growth and poverty reduction. Moreover, after many decades of neglect, tourism is no longer marginalized in economic development debates in the region or in national development planning.

As Mitchell (2007) asserts: 'with a greater focus on the potential of tourism for poverty reduction, tourism research has begun to re-join the development mainstream after a 30-year holiday from reality'. Since 2000, there has been the appearance of a surge of new critical research on tourism and development issues in southern Africa. This new wave of tourism research in southern Africa is beginning to highlight several critical issues that need to be addressed in terms of relieving constraints on tourism as a development sector in the region. By way of conclusion, this section reviews four of the major constraints that have been identified, namely the region's poor image, expensive and limited air access, limited local benefits and strategic tourism planning (see also Hall, Chapter 3 in this book).

The first set of issues relate to the limitations upon attracting high-yield long-haul international tourism caused by Africa's poor image in terms of poverty, HIV/AIDS, corruption, natural disasters and political conflicts. Table 2.4 is drawn from the travel advisory website of the British Foreign and Commonwealth Office and illustrates the extent of the 'image problem' for tourism development in contemporary southern Africa. Although the website states that across much of the region 'most visits are trouble free', it spotlights pervasive concerns about security, crime and even terrorism threats across the region. The harshest advice is provided concerning travel to Zimbabwe, where the United Kingdom citizens are cautioned to have contingency for leaving the country at short notice.

A second critical set of constraints relate to the access of international tourists to the region. Unlike Egypt, Kenya and Uganda, which have liberalized their air transport markets with generally positive results, the countries of southern Africa continue to artificially restrict international air travel by limiting the number of flights to their cities as well as the number of airlines that are permitted to fly to them (Genesis Analytics, 2006). The consequences of these restrictions are that air travel to (and within) the region is expensive and limited, the activities of low-cost carriers are curtailed and the number of tourists who can visit the region is reduced. In South Africa, the domestic air transport market and domestic tourism have both benefited from liberalization, which occurred in the 1990s and the early 2000s. At the regional and international scale of air travel, however, the pace of air transport liberalization has been slow. Mozambique, for example, continues to protect the national airline by restricting competition on international routes. The effects of these restrictions are evident in flight costs from Johannesburg to Maputo as compared to the similar distance of Johannesburg to Durban. As a result of the protectionist policies of Mozambique 'return flights to Maputo from Johannesburg are 163% more expensive than return flights to Durban' (Genesis Analytics, 2006: 8).

A third critical set of issues concerns the problem of leakages and of limited local benefits from tourism development. In Botswana, Mbaiwa (2004, 2005a, 2005b) shows that tourism in the Okavango Delta is of an 'enclave' character. Enclave development has been linked to low local linkages, repatriation of tourism profits, domination of management positions by expatriates and generally limited local development impacts. This form of tourism development has prompted calls for the advancement

Table 2.4 Travel advisories for British citizens travelling to southern Africa, January 2008

Country	Terror threat	Safety and security	Specific advice
Botswana	Low	Most visits are trouble free	Be aware of increasing incidence of violent crime. Wildlife and livestock on roads make driving hazardous
Lesotho	Low	Most visits are trouble free	Do not walk around Maseru or drive in rural areas at night
Malawi	Low	Most visits are trouble free	Driving can be hazardous; avoid travel after dark
Mozambique	Underlying	Most visits are trouble free	Be aware of the risks of violent crime, poor road safety standards and minimal health facilities
Namibia	Low	Most visits are trouble free	If travelling along the Caprivi Strip stick to well-travelled routes
South Africa	Underlying	Most visits are trouble free	There is a high level of crime, but most occurs in townships and isolated areas away from the normal tourist destinations. Standard of driving is variable and there are many fatal accidents
Zambia	Low	Most visits are trouble free, although armed robberies and car hijackings do occur	Exercise caution when travelling in rural parts of North Western, Copperbelt, Central and Luapala provinces close to DR Congo border. Be aware of risk of landmines on borders with DRC and Mozambique
Zimbabwe	Low	You are advised to have your own contingency plan of how you would leave at short notice	There has been a general increase in the level of violent crime and a serious deterioration in the economy and infrastructure making basic services, including utilities and health services, very unreliable. Avoid all political demonstrations and rallies as there is a potential for these to turn violent. Exercise extreme caution when travelling. Public services are unreliable. We advise against backpacking and travel on public transport

Source: Travel advice by country as listed on www.fco.gov.uk for 16 January 2008

and support of 'pro-poor tourism (PPT)' initiatives, including community-based tourism, across southern Africa (Ashley, 2005; Ashley & Roe, 2002). The agenda of PPT focuses on how tourism affects the livelihoods of the poor and of how positive impacts can be enhanced through targeted sets of interventions or strategies for PPT (see Ashley *et al.*, 2000; Rogerson, 2006) (see Box 2.2). In order to realize potential gains from tourism for local communities, it is argued that 'tourism development needs to be reoriented according to the interests of local stakeholders, in particular poor people' (Forstner, 2004: 497).

Box 2.2   Tourism Benefiting the Poor: PPT

JARKKO SAARINEN

Tourism is increasingly seen as a strong contributor to poverty alleviation, and one of the recent buzzwords in tourism studies is PPT. It can be simply defined as 'a tourism that generates net benefits for the poor' (Ashley & Roe, 2002: 62). These benefits may go beyond the economic issue and include social, environmental and cultural issues. In this respect, PPT may not differ much from the other existing keywords, such as ecotourism and sustainable tourism. However, according to Holden (2007) PPT as tourism development and management approach should involve a strong emphasis on poverty reduction in developing countries while ecotourism and sustainable tourism, for example, can be also linked to various other kinds of goals and settings.

The origin of the PPT dates back to the late 1990s. Much of the research was conducted in the United Kingdom-based institutions such as Overseas Development Institute (ODI) and Department for International Development (DFID), while the majority of the studies have focused on Africa and especially the southern part of the continent. Although many scholars in the tourism studies give support to the idea of PPT and other related approaches, some of them also remind of the potential danger of overemphasis of tourism as a means of poverty reduction (see Hall, 2007; Hall & Brown, 2006; Rogerson, 2006). Therefore, tourism alone should not be viewed as the last resort in rural poverty alleviation and local development strategies in developing countries. Tourism has a good potential to deliver benefits for the poor and work as a means of poverty alleviation but, based on the critics, there is also a risk to utilize 'ethical' policy goals such as poverty reduction while actually introducing (pro-poor) tourism and related activities to new areas with structures of dependency as elsewhere in the global tourism industry.

Finally, limitations in strategic tourism planning are a further constraint on tourism development in the region. Of special concern is that tourism planning across southern Africa continues to emphasize alternative tourism. Although the numbers of tourists seeking alternative tourism is growing, the actual numbers of tourists who seek an exclusively ecotourism or cultural tourism holiday is limited relative to the sea, sun and sand leisure destinations (Rogerson, 2003c). From the positive experience of mass tourism destinations in other parts of Africa as well as in other parts of the developing world (see Mitchell & Faal, 2007), it can be argued that in order to extend the developmental impact of tourism, planners in southern Africa need to reconsider the case for nurturing mainstream sun and sand holiday resort tourism. The development of an appropriate combination of 'sun, sea and sand' and of the region's existing alternative tourism products could potentially strengthen greatly the position of southern Africa within the global tourism economy.

Acknowledgements

Thanks are due to Mrs Wendy Job, Cartographic Unit, School of Geography, Archaeology and Environmental Studies, University of the Witwatersrand, Johannesburg for the preparation of all the accompanying Figures.

References

Ashley, C. (2005) *Facilitating Pro-poor Tourism with the Private Sector: Lessons Learned from Pro-poor Tourism Pilots in Southern Africa* (Working Paper 257). London: Overseas Development Institute.

Ashley, C. and Mitchell, J. (2005) *Can Tourism Accelerate Pro-poor Growth in Africa?* London: Overseas Development Institute.

Ashley, C. and Roe, D. (2002) Making tourism work for the poor: Strategies and challenges in southern Africa. *Development Southern Africa* 19, 61–82.

Ashley, C., Boyd, C. and Goodwin, H. (2000) *Pro-poor Tourism: Putting Poverty at the Heart of the Tourism Agenda* (Natural Resources Perspectives No. 61). London: Overseas Development Institute.

Boyd, S.W. and Butler, R.W. (1996) Managing ecotourism: An opportunity spectrum approach. *Tourism Management* 17 (8), 557–566.

Budowski, G. (1976) Conservation: Tourism and environmental conflict, coexistence or symbiosis? *Environmental Conservation* 3 (1), 27–31.

Butler, R.W. (1980) The concept of a tourism area cycle of evolution: Implications for management of resources. *Canadian Geographer* 24 (1), 5–12.

Butler, R.W. (1990) Alternative tourism: Pious hope or Trojan horse? *Journal of Travel Research* 28, 50–55.

Christie, I.T. and Crompton, D.E. (2001) *Tourism in Africa* (Working Paper Series No. 12). Washington, DC: World Bank Africa Region.

Cornelissen, S. (2005) *The Global Tourism System: Governance, Development and Lessons from South Africa.* Aldershot: Ashgate.

DTI (2007) *Backpacking and Youth Travel in South Africa.* Pretoria: DTI.

Elliott, S.M. and Mann, S. (2006) *Development, Poverty and Tourism: Perspectives and Influences in Sub-Saharan Africa.* Washington, DC: George Washington Center for the Study of Globalization.

Euromonitor International (2007a) *Travel and Tourism in Malawi.* London: Euromonitor International.

Euromonitor International (2007b) *Travel and Tourism in Zimbabwe.* London: Euromonitor International.

Euromonitor International (2007c) *Travel and Tourism in Swaziland.* London: Euromonitor International.

Euromonitor International (2007d) *Travel and Tourism in South Africa.* London: Euromonitor International.

FIAS and OECD Development Centre (2006) *The Tourism Sector in Mozambique: A Value Chain Analysis Volumes 1 and 2.* Washington, DC: International Finance Corporation.

Forstner, K. (2004) Community ventures and access to markets: The role of intermediaries in marketing rural tourism products. *Development Policy Review* 22, 497–514.

Genesis Analytics (2006) *Clear Skies Over Southern Africa: The Importance of Air Transport Liberalisation for Shared Economic Growth.* Report prepared for The ComMark Trust, Johannesburg.

Ghimire, K.B. (2001a) Regional tourism and South–South economic cooperation. *Geographical Journal* 167 (1), 99–110.

Ghimire, K.B. (2001b) The growth of national and regional tourism in developing countries: An overview. In K. Ghimire (ed.) *The Native Tourist: Mass Tourism Within Developing Countries* (pp. 1–29). London: Earthscan.

Gladstone, D.L. (2005) *From Pilgrimage to Package Tour: Travel and Tourism in the Third World.* Abingdon: Taylor & Francis.

Hall, C.M. (2007) Pro-poor tourism: Do 'tourism exchanges benefit primarily the countries of the South'? *Current Issues in Tourism* 10 (2/3), 111–118.

Hall, D. and Brown, F. (eds) (2006) *Tourism and Welfare: Ethics, Responsibility and Sustained Well-being.* Wallingford: CABI.

Hall, C.M. and Page S.J. (1999) *The Geography of Tourism and Recreation; Environment, Place and Space.* New York: Routledge.

Heath, R. (1986) The national survey of outdoor recreation in Zimbabwe. *Zambezia* XII(1), 26–36.

Henderson, J.C. (2006) Tourism in Dubai: Overcoming barriers to destination development. *International Journal of Tourism Research* 8, 87–99.

Holden, A. (2007) *Environment and Tourism.* London: Routledge.

Hoole, A. (2001) Integrated resource assessment and environmental management guidelines for the Zambezi River above the Victoria Falls. Environmental capacity enhancement and master plan project: Combination Plan Preparation Authority, Canadian International Development Agency, Government of Zimbabwe – Department of Physical Planning, The Victoria Falls Consortium, Harare.

Kiambo, W. (2005) The emerging role of tourism in Mozambique's post-war reconstruction. *Africa Insight* 35 (4), 142–148.

Manwa, H. (2007) Is Zimbabwe ready to venture into the cultural tourism market? *Development Southern Africa* 24 (3), 465–474.

Mashinini, V, (2003) Tourism policies and strategies in Lesotho: A critical appraisal. *Africa Insight* 33 (1/2), 87–92.

Masocha, M. (2006) Informal waste harvesting in Victoria Falls town, Zimbabwe: Socio-economic benefits. *Habitat International* 30 (4), 838–848.

Matose, F., Mudhera, M. and Mushore, P. (1997) *The Woodcraft Industry of The Bulawayo-Victoria Falls Road* (IES Working Paper 2). Harare: Institute of Environmental Science, University of Zimbabwe.

Mbaiwa, J.E. (2003) The socio-economic and environmental impacts of tourism development in the Okavango Delta, north-western Botswana. *Journal of Arid Environments* 31 (2), 447–467.

Mbaiwa, J.E. (2004) The socio-cultural impacts of tourism development in the Okavango Delta, Botswana. *Journal of Tourism and Cultural Change* 2 (2), 163–184.

Mbaiwa, J.E. (2005a) Enclave tourism and its socio-economic impacts in the Okavango Delta, Botswana. *Tourism Management* 25 (2), 157–172.

Mbaiwa, J.E. (2005b) The problems and prospects of sustainable tourism development in the Okavango Delta, Botswana. *Journal of Sustainable Tourism* 13 (2), 203–227.

Mbaiwa, J.E., Toteng, E.N. and Moswete, N. (2007) Problems and prospects for the development of urban tourism in Gaborone and Maun, Botswana. *Development Southern Africa* 24 (3), 725–739.

Mitchell, J. (2007) Fighting poverty on a sun-bed. In *Overseas Development Institute Annual Report 2007* (p. 15). London: Overseas Development Institute.

Mitchell, J. and Ashley, C. (2006). *Can Tourism Help Reduce Poverty in Africa?* London: Overseas Development Institute.

Mitchell, J. and Faal, J. (2007) Holiday package tourism and the poor in the Gambia. *Development Southern Africa* 24 (2), 445–464.

Naude, W.A. and Saayman, A. (2004) The determinants of tourist arrivals in Africa: A panel data regression analysis. Paper prepared for the International Conference, Centre for Study of African Economies, St Catherine's College, University of Oxford, 21–22 March.

RETOSA (2008) Retosa – the ultimate travel guide to Southern Africa. On WWW at http://www.retosa.co.za. Accessed 16.1.2008.

Rogerson, C.M. (2002) Urban tourism in the developing world: The case of Johannesburg. *Development Southern Africa* 19 (1), 169–190.

Rogerson, C.M. (2003a) Developing Zambia's tourism economy: Planning for 'the real Africa'. *Africa Insight* 33 (1/2), 48–54.

Rogerson, C.M. (2003b) Changing casino tourism in South Africa. *Africa Insight* 33 (1/2), 142–149.

Rogerson, C.M. (2003c) The OUZIT initiative: Re-positioning Southern Africa in global tourism. *Africa Insight* 33 (1/2), 33–35.

Rogerson, C.M. (2004) Regional tourism in South Africa: A case of 'mass tourism of the South'. *GeoJournal* 60 (3), 229–237.

Rogerson, C.M. (2006) Pro-poor local economic development in South Africa: The role of pro-poor tourism. *Local Environment* 11 (1), 37–60.

Rogerson, C.M. (2007a) Reviewing Africa in the global tourism economy. *Development Southern Africa* 24 (3), 361–379.

Rogerson, C.M. (2007b) Backpacker tourism in South Africa: Challenges and strategic opportunities. *South African Geographical Journal* 89 (1), 161–171.

Rogerson, C.M. (2007c) The challenges of developing backpacker tourism in South Africa: An enterprise perspective. *Development Southern Africa* 24 (3), 425–444.

Rogerson, C.M. and Kiambo, W. (2007) The growth and promotion of regional tourism in the developing world: The South African experience. *Development Southern Africa* 24 (3), 505–521.

Rogerson, C.M. and Lisa, Z. (2005) 'Sho't left': Promoting domestic tourism in South Africa. *Urban Forum* 16 (2/3), 88–111.

Rogerson, C.M. and Visser, G. (eds) (2004) *Tourism and Development Issues in Contemporary South Africa*. Pretoria: Africa Institute of South Africa.

Rogerson, C.M. and Visser, G. (2006) International tourist flows and urban tourism in South Africa. *Urban Forum* 17, 199–213.

Rogerson, C.M. and Visser, G. (eds) (2007) *Urban Tourism in the Developing World: The South African Experience.* New Brunswick, NJ and London: Transaction Publishers.

Saayman, M. and Saayman, A. (2003) International and African markets tourism markets for South Africa: An economic analysis. *Africa Insight* 33 (1/2), 93–98.

Sanyanga, R.A. and Musundire, H.M. (1999) Waste management in the major population centres of the Zambezi Valley – Botswana, Zambia and Zimbabwe. In N.T. Yap (ed.) *Cleaner Production and Consumption Opportunities in East and Southern Africa* (pp. 1–16). Harare: Weaver Press.

Swaziland Tourism Authority (2008) Welcome to the Kingdom. On WWW at http://www.welcometoswaziland.com. Accessed 16.1.2008.

Telfer, D.J. and Sharpley, R. (2008) *Tourism and Development in the Developing World.* London: Routledge.

Thomas, R., Pigozzi, B.R. and Sambrook, R.A. (2005) The carrying capacity measures: Crowding syndrome in the Caribbean. *The Professional Geographer* 57 (1), 13–20.

UNEP (2002): *Economic Impacts of Tourism.* On WWW at http://www.uneptie.org/pc/tourism/sust-tourism/economic.htm. Accessed 26.11.2007.

UNWTO (2006) *Tourism Highlights 2006 Edition.* Madrid: UNWTO.

Visser, G. (2002) Gay tourism in South Africa: Issues from the Cape Town experience. *Urban Forum* 13 (1), 85–94.

Visser, G. (2003a) Gay men, tourism and urban space: Reflections on Africa's gay capital. *Tourism Geographies* 5 (1), 168–189.

Visser, G. (2003b) Gay men, leisure space and South African cities: The case of Cape Town. *Geoforum* 34 (1), 123–137.

Visser, I. (2006) Approaches for SME successes. Paper presented at the UNWTO Seminar for Tourism Destination Management: Routes to Success, 27–29 March.

World Bank (2006) *Tourism: An Opportunity to Unleash Shared Growth in Africa* (Note 16). Washington, DC: Africa Private Sector Development.

ZTA (Zimbabwe Tourism Authority) (2005) *Tourism Economic Indicators 2000–2004.* Harare: ZTA.

Chapter 3

Tourism Policy and Politics in Southern Africa

C. MICHAEL HALL

Introduction

Tourism and politics have been substantially explicitly related to each other in southern Africa for many years. These relationships occur as a result of the effect that governments and their actions have had on the images of countries in the region and the region as a whole; the extent to which tourism is used as a means of economic and social development; and the way in which tourism is enmeshed within broader political structures and processes.

This chapter seeks to provide an introduction to tourism policy and politics in southern Africa and is divided into several sections. First, the chapter provides a general introduction to the literature on tourism politics and policy. Second, it describes the multi-scalar by which tourism is governed in the region. Third, it discusses the way in which political activity has influenced tourism and tourist flows and patterns. Finally, the chapter briefly outlines some of the potential future issues that tourism in the region faces in political terms.

The Tourism Policy Field

Politics and public policy are extremely significant aspects of tourism, whether they are local, provincial, national, supranational or global in scale (or constituted in terms of multi-level relationships and governance between these various scales), because of the role of state and quasi-state institutions in influencing and regulating the tourism industry and tourist activity. Public policy is a consequence of a number of factors, including political environment, values and ideologies, distribution of power, institutional frameworks and decision-making processes within different state regulatory systems. Nevertheless, as Hall and Jenkins (2004) identify 'there is little agreement in public policy studies as to what public policy is, how to identify it, and how to clarify it'. As Cunningham (1963: 229) suggested 'policy is like the elephant – you recognise it when you see it but cannot easily define it'. However, a common element in definitions is that 'public policies stem from governments or public authorities ... A policy is deemed a public policy not by virtue of its impact on the public, but by virtue of its source' (Pal, 1992: 3).

Dye (1992: 2) defined public policy as 'whatever governments choose to do or not to do'. Utilising this approach, Hall and Jenkins (1995) described tourism public policy as whatever governments choose to do or not to do with respect to tourism. Such state actions with respect to tourism are justified from a number of economic and political rationales, including

- improving economic competitiveness;
- amending property rights;
- enabling government decision makers to take account of economic, environmental and social externalities;
- providing widely available public benefits;
- reducing risk and uncertainty for investors;
- supporting projects with high capital costs and involving new technologies;
- encouraging social and economic development in marginal and peripheral areas;
- assisting specifically targeted and often marginal populations; and
- educating and providing information.

In southern Africa, all these rationales have been used at various times with respect to justifying government involvement in tourism. However, what is regarded as appropriate in terms of state intervention in tourism is not a constant but is affected by changing political ideologies (Hall & Jenkins, 2004). This has particularly been the case in southern Africa where there are often significant differences in ideological and philosophical perspectives over the appropriate role of the state with respect to ownership of productive assets and the regulation of the private sector (see Box 3.1). Many countries in the region, as well as political parties, have a substantial history of state ownership that is, at times, at odds with some of the more free market and deregulated approaches towards the role of the private sector in tourism. Indeed, tourism is often regarded as integral to empowerment of previously disadvantaged groups in a number of locations in the region (e.g. Brennan & Allen, 2001; Rogerson, 2004, see also Rogerson, Chapter 2 and Manwa, Chapter 4 in this book). For example, the African National Congress (ANC) regard tourism as an important mechanism to generate employment with their 2004 report on *Delivery to Women* stating

The Tourism Policy is also enabling, so that by 2010, more than 174,000 new jobs can be created directly in the travel and tourism industry, and 516,000 jobs can be created, directly and indirectly, across the broader South African economy. These will involve high levels of training, pay higher than average wages and be particularly accessible to women, unskilled people and new entrants to the job market. Most of the new jobs will be in areas where structural unemployment is most high. (ANC, 2004)

In many Western countries, the role of government in tourism has undergone a substantial shift from a public administration model, which sought to implement government policy for a perceived public good, to a corporatist model that emphasises efficiency, investment returns, the role of the market and relations with stakeholders, usually defined as the industry (Hall & Jenkins, 2004). Such a neoliberal perspective has had substantial influence with respect to many significant aid donors to the region, such as the World Bank, but sits uneasily with the postcolonial political agenda of many political parties in the region, which have historically been suspicious of 'market forces' as a means by which poor majorities have not been able to share some of the wealth enjoyed by colonial or indigenous elites. Therefore, it is perhaps not surprising that many parties, for example the ANC, in South Africa regard the state as the major institutional instrument by which an improved redistribution of wealth and resource access can occur.

Box 3.1   Botswana Tourism Policy: Diversifying 'Low-Volume–High-Value' Development Strategy

JARKKO SAARINEN

Botswana's first, and still current, official tourism policy was formulated in 1990. It was issued for three main reasons: (1) the tourism industry was not fully recognised in terms of government policies and priorities; (2) to capitalise the perceived growth potential of tourism at the time and (3) the Batswana were not seen likely to benefit from that potential without a new policy framework emphasising the needs of local business and people to participate in the development of the growth industry. The main objective of the policy was to gain from the abundant tourism resources of the country on a sustainable basis with the greatest possible social and economic net benefits for the Batswana. To substantially increase the financial returns from tourism to people in Botswana, a shift away from those tourists 'who are casual campers towards those occupying relatively permanent accommodation' was set as the main subordinate objective (Tourism Policy, 1990: 14). The point was illustrated in the policy paper:

> If there were a large number of regional tourists who brought with them all their provisions, camped on public grounds for which no charge was levied, purchased no services and paid no local taxes, their presence would simply impose costs arising from overcrowding of public facilities and degradation of the environment. The same would hold for overseas tourists ... if such tourists were to travel on foreign carriers, stay at facilities located on large tracts of land which only nominal rents were being charged, and pay foreign owners for services obtained in such a way that incomes did not accrue to residents and tax revenues did not accrue to government. In these circumstances, the land and wildlife 'used' by such tourists might just as well be ceded to another country. (Tourism Policy, 1990: 2)

The reading of the policy paper should be placed in a postcolonial context with a strong influence and visibility of South African self-driving and self-serviced tourists in the country. Although the global and regional tourism environment has changed considerably since the 1990s and the importance of tourism has grown from 2.5% (1989) to 3.4% (2006) of the direct contribution to the gross domestic product (GDP) in Botswana (Tourism Policy, 1990; UNWTO, 2008), the underlining problem has not disappeared: how to increase and expand the positive impacts of tourism to the local people and regional economies.

In 2008, the original policy document was under revision. This revised tourism policy for Botswana was outlined by UNWTO (2008) aiming to gear the country's tourism development to growth. According to UNWTO it is clear that, unlike the neighbouring South Africa and Namibia, the country has not been adequately capitalised on the positive international tourism growth in the region. Therefore, Botswana should take urgent steps to defend and increase its market share (UNWTO, 2008: 2). A vision of the policy is set for 2020 when Botswana should be globally renowned as the most authentic and exiting wilderness tourism destination in the world with large numbers of the people of Botswana participating in, and benefiting from, the industry. The key goals that aim to achieve the vision are

- to elevate the recognition of tourism as a priority growth sector in Botswana;
- to increase the contribution of tourism to the GDP;

- to achieve exceptional growth in tourism volumes, lengths of visitor days and expenditure;
- to substantially increase the share of oval ownership and management in the industry;
- to promote labour-intensive tourism practices;
- to encourage growth in tourism entrepreneurship;
- to advance, promote and support investments and linkages in tourism;
- to encourage community participation;
- to create awareness of tourism among the population;
- to develop and improve tourism skills and provide tourism education and training;
- to ensure easy access for prospective tourists to and within the country;
- to ensure a safe and secure tourism environment;
- to encourage a tourism culture among the Botswana; and
- to become a globally known leader in environmental tourism and nature conservation.

The new policy recognises that the current position of Botswana as 'low-volume–high-value' (i.e. limited tourist numbers with high expenses) destination offers too narrow opportunities to develop the industry. Hence, 'there is a need to expand the positioning to broaden the range of middle-to-high tourism market segments and products' (UNWTO, 2008: 5). The policy also notes that the most of the valuable resources for exclusive safari lodge business have already been developed, which reduces growth opportunities within the current policy framework from 1990. However, the attraction elements of the country have remained the same in both the policies. The main emphasis on the first policy paper was targeted to wildlife and wilderness experiences (Tourism Policy, 1990: 1), whereas the revised document wraps up the vision to wilderness tourism destination with a slogan 'to be touched by wilderness' (UNWTO, 2008: 9). Similarly both the documents also side mention the role of culture as an additional potential value for tourism, but the emphasis to diversify the product is more evident, although in a minimal scale, in the new tourism policy for Botswana. For example, the new policy refers to a need to develop man-made tourist attractions in and around the capital city of Gaborone and other major towns, but it also emphasises the role of local people in a more general level by stating that the future tourism development in Botswana should be government led, private sector driven and community-based (UNWTO, 2008: 10).

The first tourism policy paper for Botswana aimed to guide the tourism industry in order to avoid the negative consequences of mass scale nature-based tourism whose impacts were evident at the time in East Africa, for example, by setting the 'low-volume–high-value' rule in practice. The nature conservation point of view has most probably been very successful but interestingly it did not achieve the societal targets it actually aimed for: to benefit the local people and economies based on the full potential of the industry. This is not to say that tourism has not profited the nation and local people in the form of salaries and employment (see Mbaiwa & Darkoh, in this volume), for example, but by setting the elite 'safari landscape' focusing mainly on international overseas tourists, the expectations concerning the quality of tourist facilities, products and service, among other things, have reduced the possibilities of local people to be actively involved with the industry at all levels, including ownership, entrepreneurship and managerial positions. On the other hand, this kind of policy emphasis has perhaps assisted international investors, entrepreneurs, tour companies, managers and expatriate staff to enter the country's tourism system. As a result and manifestation of this, the estimated leakage of tourism revenue in Botswana is over 70%. In the Okavango Delta and Maun, which is the hot spot area of tourism in the country, only 18.5% of the tourism businesses are purely citizen-owned, whereas non-citizens owned 53.8% of the tourism enterprises in 2001 (Mbaiwa, 2005). According to Mbaiwa, the average charge for accommodation per night in the Delta area in 1999 was US$400, the pricing level of which calls for facilities, capital and business knowledge that are not usually embedded in peripheries (see Hall, 2005).

It seems that the aim to 'get rid of a happy camper' did not solve the problem of the low level of local benefits from tourism, and thus the same challenge forms a basis for the new policy framework which aims to broader the visitor segments and encourage domestic tourism. This will most probably increase the tourist flows in the country, but by focusing still almost only on wildlife and wilderness-based tourism there is a danger of neglecting the local communities in tourism development and also planning in the future. The new tourism policy framework for Botswana positively highlights the needs, participation and benefits of local communities, but their role in the country's tourism brand, image and products is still minor or non-existent. However, in cultural and community-based tourism, in contrast to wildlife viewing in safari tourism and so on, the 'outsourcing' of the local people with their knowledge, traditions and heritage is perhaps less likely to happen, although the forms of culture tourism are not free either from the neoliberal nature or from the influences of the global tourism industry. Still, cultural and community-based tourism could be used more to complement the present and emphasised wilderness tourism.

Interestingly, the growth targets (indicators) of the new policy, which are based on the key goals listed above, do not refer any community involvement or local ownership-related goals but on tourism and its indicators. However, the focus of the planned new tourism policy is rhetorically set to serve the Batswana with the key principle referring to community-based tourism development, but by emphasising solely the tourism-related indications and targets the industry becomes an end in the development process. Peter Burns (1999) points this outcome with the term 'tourism first', in contrast to the term 'development first' (referring to the idea of tourism as a tool instead of an end), in which the tourism industry and its needs dominate the development discourses and practices. This potentially challenges the connection between tourism and sustainability, because the desired goals of community development are not necessarily the same as the tourism industry's goals (see Butler, 1999; Mbaiwa, 2003, 2004, 2005), whose contradiction may lead to 'low benefits–high costs' for the local people.

The definition of tourism public policy used by Hall and Jenkins (1995) covers government action, inaction, decisions and non-decisions, and implies a deliberate choice between potential policy alternatives. Such an approach regards tourism policy as a process that involves the formulation and implementation of tourism-related policy in dynamic environments where there is a complex pattern of decisions, actions, interaction, reaction and feedback (Hall & Jenkins, 2004).

For a policy to be regarded as public policy, at the very least it must have been processed, even if only authorised or ratified, by public agencies (Hall & Jenkins, 1995). However, it should be emphasised that in the case of tourism, relevant public policies that affect tourism may not be generated by tourism-specific agencies or ministries. In fact, Hall (2008) argues that the policies established by non-tourism government bodies, for example, with respect to visa and entry policy, tax, labour law, business regulation, foreign investment, transport and aviation, land-use planning and environment, are likely to have much more impact on tourism than the policies that are specifically entitled 'tourism' (Figure 3.1) and that usually focus on marketing and, to a limited extent, tourism-specific investment and development issues. In addition, tourism public policy even if processed by public agencies 'may not have been significantly developed within the framework of government' (Hogwood & Gunn,

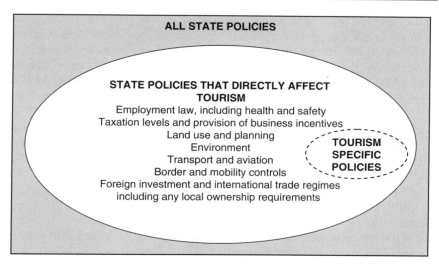

Figure 3.1   Tourism policy fields

1984: 23). Therefore, besides seeking to analyse the tourism-related actions and inactions of governments, it is also important to understand the role of other elements of the policy process, including

- interest groups (e.g. tourism business and industry associations, conservation groups, trade unions and community groups);
- significant individuals (e.g. business leaders, tribal and community leaders);
- political parties;
- members of the public service (e.g. employees within government bodies including members of tourism organisations or regional economic and social development agencies);
- other institutions (e.g. other levels of governance as well as the organisations that operate within them); and
- other individuals (e.g. academics, researchers and consultants),

all of which influence and perceive public policies that affect tourism in significant and often markedly different ways (Hall, 2008; Hall & Jenkins, 1995, 2004). Furthermore, in examining tourism policy and planning, outputs (a policy statement or plan) need to be distinguished from outcomes (the actual effects of policies and plans), as policy outcomes may lead to quite different results from those that were originally intended (Hall, 2008).

In the South African case, for example, although tourism is clearly a major focus of government activities, there are clearly substantial variations in the extent of policy focus on tourism between key policy actors. In South Africa, as with countries such as Australia, Canada and New Zealand (Hall, 2008), an organisational split has been made at the national level with marketing the responsibility of South African Tourism and policy, and administration of the Department of Environmental Affairs and Tourism. In addition, there are also government agencies responsible for education and training (Tourism, Hospitality, Education and Sports Training Authority (THETA)), quality evaluation (Tourism Grading Council) and national parks (South Africa National Parks).

Tourism is one of the major foreign exchange earners for South Africa and is regarded as an important employment generator because of the relatively low skill base that is required. South Africa also expects to attract 10 million foreign tourists by 2010 when the country hosts the FIFA World Cup (Parliament of the Republic of South Africa, 2007). However, no member of the tripartite governing alliance between the ANC, the South African Communist Party and the Congress of South African Trade Unions has a separate tourism policy, although tourism can be identified as an element in parliamentary submissions, speeches and other organisational documents. The ANC, in particular, appear to link government tourism policy documentation and organisation, particularly with respect to bringing tourism and environment together under a single ministry as synonymous with party policy. In contrast, other political parties, such as the Democratic Alliance (DA) and the Inkatha Freedom Party (IFP), have separate tourism policy documents.

Although political party statements with respect to tourism may be quite general in style, they can contain significant differences with respect to tourism administration and policy. For example, the tourism policy of the IFP (nd) provides some relatively concrete policy measures and states that 'there may be some benefit in delinking tourism from environmental affairs and either establishing a separate ministry, or linking tourism with a mainstream ministry such as Trade and Industry' as well as also suggesting that with respect to incentives for the tourism industry 'the IFP believe that tourism enterprises should be given the same status as manufacturing enterprises'.

The policy paper of the DA is almost virtually interchangeable with that of many free-market-oriented political parties in Western democracies noting that tourism's potential to contribute to economic growth, job creation and foreign exchange earnings is enormous. To remain one of the world's fastest-growing tourist destinations, South Africa must be competitive and safe, ensure that seats are available on aeroplanes and properly coordinate governmental support.

To grow the tourism market, the DA (nd) will

- ensure close cooperation in promoting tourism between the state and the private sector;
- encourage greater involvement by private entrepreneurs;

- encourage innovative product development;
- make service excellence a key objective;
- diversify the product attractions;
- increase funding to promote South African tourism;
- promote South Africa as a unique brand;
- liberalise airline services to and from South Africa;
- minimise red tape in the issuing of visas;
- improve public transport; and
- facilitate the provision of training.

In addition to the influence of political parties on government tourism policies and agencies, industry organisations are also an important element in policy formulation and implementation. In South Africa, as with many developed countries, there is a lead tourism business association that acts as an 'umbrella group' for the broad interests of the private sector (The Tourism Business Council of South Africa (TBCSA) as well as a series of representative associations of tourism and hospitality sub-sectors (Association of South African Travel Agents (ASATA), Backpacking South Africa (BSA), The Bed and Breakfast Association of South Africa (BABASA), Board of Airline Representatives of South Africa (BARSA), Coach Operators Association of Southern Africa (COASA), Federated Hospitality Association of South Africa (FEDHASA), Field Guides Association of Southern Africa (FGASA), Guest House Association of Southern Africa (GHASA), Restaurant Association of South Africa (RASA), Southern African Association for the Conference Industry (SAACI), Southern Africa Tourism Services Association (SATSA) and Southern African Vehicle Rental and Leasing Association (SAVRALA)).

Such a proliferation of tourism-related organisations – along with the interests of provincial and regional government and their agencies – clearly makes for a congested policy arena. However, there is a substantial degree of cooperation and partnership between the associations, particularly the TBCSA, and government, with the TBCSA being represented on key government tourism bodies. While the strength of public–private partnerships may be seen by some as a good thing in that it encourages 'coordination', such a situation can be described in policy terms as a 'sub-government' in which policy alternatives can be limited to the relatively narrow input of a small selection of institutional and individual actors, which means that many policy disagreements can be internalised within the sub-government rather than being part of broader policy debates or discussion on a wider selection of policy alternatives. As Tetekin (2003) observed, in South Africa tourism and politics are inseparable. Yet in some cases sub-governments can also lead to situations in which small tourism businesses and micro-entrepreneurs may find themselves at a disadvantage in policy debates relative to larger firms that tend to have their interests more strongly represented in sub-government structures (Hall, 2008).

However, the extent to which this situation may exist in the South African context or in other states in the region has not been researched.

Unfortunately, there is little specific policy analysis of tourism policy processes in southern Africa with the primary focus instead being on particular dimensions such as economic development initiatives, especially with respect to pro-poor tourism (PPT) and ecotourism-related ventures (e.g. Ashley & Roe, 2002; Binns & Nel, 2002; Brennan & Allen, 2001; Manwa, 2003; Rogerson, 2002a, 2002b, 2003a, 2003b, 2005, 2006; Rogerson, Chapter 2 in this book; Rogerson & Visser, 2004; Visser & Rogerson, 2004). Perhaps just as significantly, many policy documents or papers assessing the potential contribution that tourism can make to economic and social development in southern Africa, or in the continent as a whole, often fail to explicitly discuss the political context of tourism policy-making and development, especially with respect to power relationships, the role of different actors and institutions, and the realities of decision-making processes (e.g. Aylward & Lutz, 2003; Cornelissen, 2005; Dieke, 2000, 2003). Such a situation therefore highlights the need for more detailed and sophisticated analysis of tourism policy making in the region, especially given its multi-level nature.

Multiple Level Governance of Tourism in Southern Africa

The governance of tourism is increasingly being described in terms of the characteristics of multi-level governance (Hall, 2008). Governance is 'the capacity for steering, shaping, and managing, yet leading the impact of transnational flows and relations in a given issue area, through the inter-connectedness of different polities and their institutions in which power, authority and legitimacy are shared' (Morales-Moreno, 2004: 108–109). Tourism is a policy area that bears the hallmarks of being maintained not just by territorial state-bounded authorities, but 'by a network of flows of information, power and resources from the local to the regional and multilateral levels and the other way around' (Morales-Moreno, 2004: 108). This is especially the case in southern Africa where tourism in the region is substantially integrated into the international political economy of tourism.

Figure 3.2 attempts to illustrate the nature of multi-layered tourism governance. The model indicates that, as noted above, tourism policy is just one policy area that lies at the intersection of a number of policy areas that affect tourism, and which occur at various levels of governance. Any specific tourism policy arena, that is, the specific configuration of institutional and individual actors, their institutional arrangements, values and power relations that relate to specific policy issues therefore occurs over multiple scales and may involve actors over a number of policy areas. However, different institutions and organisations will have greater political and

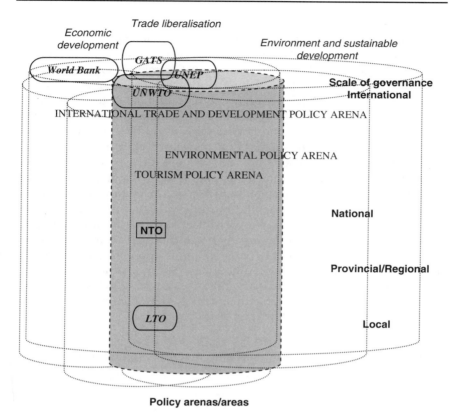

Figure 3.2 Multi-layered tourism governance

regulatory influence at some scales than others but will often act over multiple scales. The figure indicates international trade and economic development, trade liberalisation and environment, and sustainable development as just three of the policy areas that intersect with tourism policy making and that will impact on tourism development at the local level, even though many tourism businesses at this scale may be completely unaware of the potential impacts of decision-making at the international level may have on them. For example, inclusion of tourism within a General Agreement on Trade in Services (GATS) would potentially have substantial effects on such matters as foreign investment in tourism services, tourism marketing, mobility of foreign workers and transport (Coles & Hall, 2008). In fact, trade liberalisation in services is a good example of where tourism is a subset of broader policy-making concerns and where trade in international tourism can only really be understood if a wider public policy perspective is adopted.

Therefore, critical to much tourism development in southern Africa is the role of international institutions and policy actors that may be supranational organisations or a national development aid agency. For example, Table 3.1 outlines tourism-related donor assistance in Mozambique for various international actors as of 2005. Indeed, international actors, including the World Bank, UNEP and UNWTO, are highly significant in all of the countries of the region and have substantially influenced tourism policy making through their policy and financial support of ecotourism and PPT initiatives as well as the involvement in tourism master plans. However, there has been little evaluation of the value or success of such plans with respect to their implementation and the extent to which they contribute to the goal achievement.

At a supranational level, the Southern African Development Community (SADC) is also increasingly significant for tourism in the region (also see Manwa, Chapter 4 in this book), particularly as a result of encouraging greater tourist mobility within member countries via a common visa initiative as well as the Regional Tourism Organization of Southern Africa, the official SADC tourism body. Although the South African supranational regime and associated multi-lateral initiatives are of growing importance for tourism, national political actions and policies continue to have the greatest impact on tourist flows and patterns.

Table 3.1 International actors in tourism-related donor assistance in Mozambique

International governmental organisations	*National governmental organisations*	*International non-governmental organisations*
• World Bank	• USAID	• Peace Park Foundation
• International Finance Corporation (member World Bank Group	• SIDA (Sweden)	• IUCN • SNV (Dutch development NGO)
• Commonwealth	• GTZ (German Development Cooperation)	• TechnoServe (US NGO)
• European Union	• KfW (German Investment and Development Bank)	• Ford Foundation
	• Irish Aid	
	• DANIDA (Denmark)	
	• AFD (France)	
	• AusAID (Australia)	

Source: Derived from International Finance Corporation (2005)

The Influence of Politics and Political Instability on Tourist Flows

Perceptions of political stability and security have significant impacts on tourist flows. In the international arena, national security concerns have not only influenced travel behaviour but have also affected the direct personal security measures employed to protect the travelling public (Hall & Jenkins, 2004). For example, the 9/11 attacks on the United States and the escalation of conflict in the Middle Eastern are regarded as having led to a reassessment of the relative safety of South Africa by Western tourists with extremely positive results in terms of increased arrivals (Tetekin, 2003). However, critical to such a change in perspective was the role of the media in representing South Africa rather than any intrinsic change in relative tourist risk in the country. Therefore, as not only is the media significant in terms of the images that surround travel and specific destinations and which influence travel decision-making, but the media also has a potentially substantial impact on the policy measures which governments take with respect to tourists safety and security (Hall, 2002).

Arguably, the influence on national politics on tourist flows, as well as the potential role of the media, is indicated in the contemporary issues facing Zimbabwe and the methods used by Robert Mugabe's regime to cling to power. Machipisa (2001) noted that invasions of commercial farms by war veterans, violence in the run-up to parliamentary elections and the harassment of tourists by ex-combatants at Victoria Falls were all given extensive international media coverage and have put Zimbabwe on the list of unsafe destination for visitors from 2000 onwards. For example, the negative perceptions of Zimbabwe around the time of the 2000 elections were reported to have led to a 60% drop in hotel occupancy, and a consequent massive retrenchment of staff (Thondlana, 2000). At that time, tourism was Zimbabwe's third largest economic sector, contributing about 6% of the country's GDP and employing 20,000 workers. Significantly, Thondlana (2000) also reported that the crisis was having a wider effect on tourism in the southern African region as a result of tourist and media perceptions of region-wide instability. However, those particular effects were then shifted by other world events. Indeed, many of the countries surrounding Zimbabwe now appear to be substantially benefiting from the political and economic crises that have affected the country (Machpisa, 2001).

Otton (2006) directly links the substantial growth in Zambian tourism with the Zimbabwe crisis noting the stark contrast between the two countries:

> The reversal of fortunes is most starkly illustrated at the Victoria Falls, one of the world's great tourist attractions that straddles the common border. In the nearby Zambian town of Livingstone, named after the

British explorer David Livingstone who was the first Westerner to set eyes on the falls, new hotels and a first class shopping mall have sprung up in recent months, while work to extend the runway at the local airport is currently ongoing. While the once sleepy Livingstone is bustling with activity, cut-price deals on the Zimbabwean side have failed to fill the vacancies in hotels such as The Royal, a traditional byword for luxury.

However, it should be noted that although South African tourism has benefited from Zimbabwe's political and economic crisis, it is not immune from concerns about political issues in southern Africa. For example, in a report on travel and tourism in South Africa, Euromonitor International (2008) reported that there was substantial concern in the South African tourism industry with respect to a decline of almost 14% in the number of inbound tourists from China, with many groups cancelling their bookings over concerns relating to crime and security in South Africa.

Presidential elections in December 2005 accompanied with massive protests created significant political and civil unrest that was well publicized through mass media. In light of this, South African Tourism aimed to counter such perceptions through awareness and marketing campaigns, while encouraging more balanced media reporting on South African society. Arrivals from China grew by 6% in 2006. (Euromonitor International, 2008: Executive summary)

Future Issues

This chapter has provided a brief outline of some of the significant tourism policy concerns facing southern Africa. Arguably, one of the most important is the sheer need for policy analysis that goes beyond superficial description to one that seeks to make transparent the interaction between various actions and the decision-making processes they are involved in. From a political perspective this means not only analysing who are the winners and losers in tourism policy making but also why. Unfortunately, such a situation can be extremely difficult in some countries where critical analysis may be taken as a criticism of government and therefore leads to a cutting off of funds, termination of contracts or worse. The potential for such actions has been put forward as one of the reasons for the lack of attention for a critical research focus on tourism policy and politics (Hall, 1994). Indeed, although there has been analysis of some of the political dimensions associated with the increased integration of the region in the international tourism system and the associated set of political–economic relationships (e.g. Cornelissen, 2005), there has been a dearth of such analysis at the national and sub-national level, where the interplay of relations between actors and institutions actually play out in terms of planning processes, decisions and their effects.

Such concerns are not merely academic as appropriate public policy analysis can help governments achieve their development aims via providing an improved understanding of policy formulation and implementation. This may be especially important considering the substantial emphasis given to the tourism benefits of South Africa hosting the 2010 FIFA World Cup as well as the specific focus on ecotourism and PPT strategies. The latter perhaps provides an extremely important illustration of the need for appropriate analysis.

The range of opinions regarding PPT and its potential to act as a positive force for human well-being is well summed up by Hall and Brown (2006: 13), 'does PPT simply offer another route by which economic imperialism, through tourism, may extend its tentacles, or is it an appropriately liberating and remunerative option?' Similarly, Hall (2007: 117) noted, 'whether tourism is a means of reducing poverty gaps beyond isolated gaps or is perhaps a symptomatic of a causal relationship, at least with respect to the broader scope of north–south trade, needs to be debated much further than what is presently the case in tourism policy circles'. Indeed, several NGOs have been highly critical of the concept of reducing poverty via tourism. For example, Ecumenical Coalition on Tourism Executive Director Ranjan Solomon (2005) stated:

> For as long as the rich and powerful are going to draw up the parameters and architecture of tourism policy, nothing will change – not much, in any case. How can it? For after all, the investor is there to make profits. Social responsibilities do not factor – evidence of this is too thin to be counted or weighed in. The occasional burst of charity is not what we are talking about and asking for. Tourism is, virtually, for all intents and purposes, one with a purely economic function in-sofar as the industry is concerned . . . .

Nevertheless, it should be noted that governments in the region, and especially South Africa, have laid substantial emphasis on the development of domestic and intra-regional tourism as a means of creating both economic development and employment opportunities as well as forging new concepts of identity and rights. As South African President Thabo Mbeki commented at the tourism conference in 2003, 'Rest and recreation is not a luxury. Neither is it a sin punishable by a life sentence in hell or purgatory. It is a human necessity, without which we cease to be human' (Mbeki, 2003).

Aside from concerns over the appropriateness of tourism policy settings, it is likely that the relationship between perceived political instability and security issues and tourism in the region will continue to dominate the political dimensions of tourism for the foreseeable future. However, arguably the media focus on a hallmark event such as the 2010 World Cup will potentially lead to more attention being given to security as well as to issues such as human rights, while such events can also become focal points for political protest. Nevertheless, it is also likely that if expectations of increased visitor numbers in the region and associated economic and employment benefits are not met then tourism may finally begin to be subject to the policy scrutiny that it richly deserves.

References

African National Congress (2004) *Report on Delivery to Women, ANC Sub-committee on Gender Issues*. On WWW at http://www.anc.org.za/ancdocs/reports/2004/delivery_to_women/environ.html. Accessed 1.4.2008.

Ashley, C. and Roe, D. (2002) Making tourism work for the poor: Strategies and challenges in southern Africa. *Development Southern Africa* 19 (1), 61–82.

Aylward, B. and Lutz, E. (eds) (2003) *Nature Tourism, Conservation, and Development in Kwazulu-Natal, South Africa*. Washington, DC: The International Bank for Reconstruction and Development/The World Bank.

Binns, T. and Nel, E. (2002) Tourism as a local development strategy in South Africa. *The Geographical Journal* 168 (3), 235–247.

Brennan, F. and Allen, G. (2001) Community-based ecotourism, social exclusion and the changing political economy of KwaZulu-Natal, South Africa. In D. Harrison (ed.) *Tourism and the Less Developed World: Issues and Case Studies* (pp. 203–221). New York: CABI Publishing.

Burns, P. (1999) Paradoxes in planning: Tourism elitism or brutalism? *Annals of Tourism Research* 26 (2), 329–348.

Butler, R. (1999) Sustainable tourism: A state-of-the-art review. *Tourism Geographies* 1 (1), 7–25.

Coles, T. and Hall, C.M. (2008) *International Business and Tourism: Global Issues, Contemporary Interactions*. London: Routledge.

Cornelissen, S. (2005) *The Global Tourism System: Governance, Development and Lessons from South Africa*. Aldershot: Ashgate.

Cunningham, G. (1963) Policy and practice. *Public Administration* 41, 63.

Democratic Alliance (nd) *The Sky's the Limit*. Democratic Alliance Tourism Policy. On WWW at http://www.da.org.za/da/Site/Eng/Policies/Downloads/Tourism.asp. Accessed 1.4.2008.

Dieke, P.U.C. (ed.) (2000) *The Political Economy of Tourism Development in Africa*. Elmsford: Cognisant Press.

Dieke, P.U.C. (2003) Tourism in Africa's economic development: Policy implications. *Management Decision* 41 (3), 287–295.

Dye, T. (1992) *Understanding Public Policy* (7th edn). Englewood Cliffs, NJ: Prentice-Hall.

Euromonitor International (2008) *Travel and Tourism in South Africa*. London: Euromonitor International.

Hall, C.M. (1994) *Tourism and Politics*. Chichester: Wiley.

Hall, C.M. (2002) Travel safety, terrorism and the media: The significance of the issue-attention cycle. *Current Issues in Tourism* 5 (5), 458–466.

Hall, C.M. (2005) *Tourism: Rethinking the Social Science of Mobility*. Harlow: Prentice-Hall.

Hall, C.M. (2007) Pro-poor tourism: Do 'tourism exchanges benefit primarily the countries of the South'? *Current Issues in Tourism* 10 (2/3), 111–118.

Hall, C.M. (2008) *Tourism Planning* (2nd edn). Harlow: Pearson.

Hall, C.M. and Jenkins, J.M. (1995) *Tourism and Public Policy*. London: Routledge.

Hall, C.M. and Jenkins, J.M. (2004) Tourism and public policy. In A. Lew, C.M. Hall and A.M. Williams (eds) *Companion to Tourism* (pp. 525–540). Oxford: Blackwell.

Hall, D. and Brown, F. (eds) (2006) *Tourism and Welfare: Ethics, Responsibility and Sustained Well-being*. Wallingford: CABI.

Inkatha Freedom Party (nd) *Tourism Policy*. On WWW at http://www.ifp.org.za/. Accessed 1.4.2008.

Hogwood, B. and Gunn, L. (1984) *Policy Analysis for the Real World*. Oxford: Oxford University Press.

International Finance Corporation (2005) *The Tourism Sector in Mozambique: A Value Chain Analysis* (Vol. II). Washington, DC: International Finance Corporation.

Machipisa, L. (2001) Sun sets on Zimbabwe tourism. *BBC News* (World Edition, 14 March, 18:32 GMT). On WWW at http://news.bbc.co.uk/2/hi/africa/1220218.stm. Accessed 1.4.2008.

Manwa, H. (2003) Wildlife-based tourism, ecology, sustainability: A tug-of-war among competing interests in Zimbabwe. *Journal of Tourism Studies* 14 (2), 45–54.

Mbaiwa, J.E. (2003) The socio-economic and environmental impacts of tourism in the Okavango Delta, Northwestern Botswana. *Journal of Arid Environments* 54 (2), 447–468.

Mbaiwa, J.E. (2004) The success and sustainability of community-based natural resource management in the Okavango Delta, Botswana. *South African Geographical Journal* 86 (1): 44–53.

Mbaiwa, J.E. (2005) Enclave tourism and its socio-economic impacts in the Okavango Delta, Botswana. *Tourism Management* 26 (2), 157–172.

Mbeki, T. (2003) Address at the opening of the tourism Indaba, 4 May 2003. On WWW at http://www.anc.org.za/ancdocs/history/mbeki/2003/tm0504.html. Accessed 1.4.2008.

Morales-Moreno, I. (2004) Postsovereign governance in a globalizing and fragmenting world: The case of Mexico. *Review of Policy Research* 21 (1), 107–117.

Otton, C. (2006) Zimbabwe crisis signals boom time for Zambian tourism. *The Statesman* (Ghana) (4 October). On WWW at http://www.thestatesmanonline.com/pages/news_detail.php?newsid=799&section=4. Accessed 1.4.2008.

Pal, L.A. (1992) *Public Policy Analysis: An Introduction*. Scarborough: Nelson Canada.

Parliament of the Republic of South Africa (2007) *The Strategic Imperatives for South Africa as Set Out in the 2007 State of the Nation Address: An Oversight Tool for Members and Committees of Parliament*. Pretoria: Parliamentary Research Unit, Parliament of the Republic of South Africa. At http://www.parliament.gov.za/live/content.php?Item_ID=224.

Rogerson, C.M. (2002a) Tourism – a new economic driver for South Africa. In A. Lemon and C.M. Rogerson (eds) *Geography and Economy in South Africa and its Neighbours* (pp. 95–110). Aldershot: Ashgate.

Rogerson, C.M. (2002b) Driving developmental tourism in South Africa. *Africa Insight* 32 (4), 33–42.

Rogerson, C.M. (2003a) Towards pro-poor local economic development: The case for sectoral targeting in South Africa. *Urban Forum* 14 (1), 53–79.

Rogerson, C.M. (2003b) Tourism planning and the economic revitalisation of Johannesburg. *Africa Insight* 33 (1/2), 108–115.

Rogerson, C.M. (2004) Tourism, small firm development and empowerment in postapartheid South Africa. In R. Thomas (ed.) *Small Firms in Tourism: International Perspectives* (pp. 13–33). Oxford: Elsevier.

Rogerson, C.M. (2005) Unpacking tourism SMMEs in South Africa: Structure, support needs and policy response. *Development Southern Africa* 22 (5), 623–642.

Rogerson, C.M. (2006) Pro-poor local economic development in South Africa: The role of pro-poor tourism. *Local Environment* 11 (1), 37–60.

Rogerson, C.M. and Visser, G. (eds) (2004) *Tourism and Development Issues in Contemporary South Africa*. Pretoria: Africa Institute of South Africa.

Solomon, R. (2005) Tourism: A challenge for the 21st century. *eTurboNews* (September 25). On WWW at http://travelwirenews.com/eTN/27SEPT2005.htm.

Tetekin, V. (2003) South Africa: Paradise with earthly problems. *Sovetskaya Rossiya* (Moscow) (13 May). Translation from Russian. On WWW at http://www.anc.org.za/ancdocs/pubs/umrabulo/umrabulo19/paradise.html.Accessed 1.4.2008.

Thondlana, B. (2000) Zimbabwe: Negative image slays tourism. *African Business*, July–August.

Tourism Policy (1990) Government Paper no. 2 of 1990. Gaborone: The Government Printer.

UNWTO (2008) Policy for the Growth and Development of Tourism in Botswana.

UNWTO/Government of Botswana Project for the Formulation of a Tourism Policy for Botswana, July 2008. UNWTO and Department of Tourism.

Visser, G. and Rogerson, C.M. (2004) Researching the South African tourism and development nexus. *GeoJournal* 60 (3), 201–215.

Chapter 4

Women and Tourism in Southern Africa

HARETSEBE MANWA

Introduction

The formal recognition of women's issues and their incorporation into Southern African Development Community (SADC) structures and the development of a regional gender policy have taken a long time and are still ongoing. It was only in 1994 when 'Women in Development Southern Africa Awareness' (WIDSAA) was established as the gender programme of the Southern Africa Research and Documentation Centre (SRDC) as one of its specialist departments.

The objectives of WIDSAA are as follows:

(1) To empower women and thereby advance the position of women in SADC member states.
(2) To engender all structures so as to facilitate gender equity.

The SADC Declaration on Gender and Development was adopted in 1997. Because it is not yet a Protocol signed by all member states it is not legally binding. The 1997 Declaration covers the following areas:

(1) The equal representation of men and women within decision-making bodies in member states and the SADC structures.
(2) A reduction in the level of poverty among women through ensuring their access to and control over productive resources such as land, credit and formal employment.
(3) The repeal and reform of legislative and constitutional provisions that discriminate against women.
(4) The need to enact gender-sensitive laws that would empower women.
(5) The need to change socio-cultural practices that discriminate against women.

(6)   The need to enhance access to quality education for men and women and to remove gender stereotyping in the curriculum.
(7)   The provision of quality reproductive and other health services to men and women.
(8)   The protection of the human rights of women and children.
(9)   The recognition and protection of the reproductive and sexual rights of women and the girl child.
(10) The need to take urgent measures to deal with and prevent increasing levels of violence against women and children.
(11) The need to encourage the mass media to disseminate information and materials in respect of human rights of women and children.

The SADC's treaty is the cornerstone of SADC's common development agenda (SADC, 1992), which is focused on deeper regional integration and poverty eradication with the principles of balance, equity and mutual benefits among the countries. The SADC tourism activities and programmes should therefore be in line with this agenda. SADC has come up with a number of legal instruments to define the type of tourism they would like to see developed in the region. Four of these are briefly discussed below.

SADC Legal Instruments

The Regional Indicative Strategic Development Plan (RISDP) (SADC, 2002) was developed in 2002. RISDP provides for the strategic direction of activities, and programmes for developing, promoting and marketing tourism (among other sectors) in the SADC region. Tourism was identified in the RISDP within the restructuring process as a priority intervention area in the SADC Agenda. It presents the challenges that need to be met, the strategies that need to be adopted, and sets the targets that are to be met in the sector. Of importance to this chapter are the following objectives of RISDP:

(1)   To promote community–public–private partnerships (CPPPs) in tourism development including small- and medium-sized enterprises (SMEs), women and youth.
(2)   To develop the SADC Trans-Frontier Conservation Areas(TFCAs) and tourism-based Spatial Development Initiatives (SDIs) as tourism product varieties.

The RISDP also sets out a number of targets that must be achieved by member states. These range from development of tourism policy to product development and marketing. Importantly, RISDP talks about gender mainstreaming of tourism. This therefore implies that the region is aware that there is gender inequality and/or discrimination in employment, or that the industry is gendered in types of organizations and ownership of organizations that make up the tourism industry.

SADC countries also signed the Protocol on Tourism Development in 1998, which came into force in November 2002. The protocol emphasizes the promotion of SMEs, and participation of women and youth in the tourism industry. The third legal instrument used by the SADC governments is the Regional Tourism Organization of Southern Africa (RETOSA) charter, which was signed by SADC Member States in 1997 and came into force in 2000. The final legal instrument is the New Partnership for Africa's Development, which has identified tourism as an important vehicle in addressing the current challenges facing the African continent. The legal instruments show SADC countries' commitment and belief in tourism being an answer to achieving the Millennium Development Goals (MDGs) of eliminating poverty in the region (RETOSA, 1997; RISDP, 2002; SADC, 1998).

The chapter is based on policy documents and reports from newspapers, research done by the author in three southern African countries (Botswana, Lesotho and Zimbabwe) and a literature review. The chapter has adopted Hall's (1999) gender-aware framework as well as Williams' (2002) framework to discuss women and tourism in southern Africa. Hall's framework argues that women's position in tourism should not be divorced from their general position in society where they have prescribed gender roles. Similar roles are transferred to tourism organizations. Williams' (2002: 5) framework emphasizes the impacts of tourism on women and classifies them into four categories: employment in the formal labour market; informal sector and sustainable livelihoods; women's social and economic empowerment; and women's influence and decision-making around tourism development. The rest of the chapter is therefore structured as follows: after the introduction, Section 2 presents a limited review of available literature on gender and tourism to contextualize the southern African experiences, Section 3 presents case studies from southern Africa and the final section of the paper concludes the issues.

Informal Sector and Sustainable Livelihoods

Tourism has been acclaimed to be the world's largest industry (Brohman, 1996; Cattarinich, 2001; Ghimire, 1997; Goodwin, 2000; Mowforth & Munt, 1998), which can be used to alleviate poverty, especially in marginalized areas, most of which are in rural parts of the developing world (Dao, 2004). The development of tourism is in line with the MDG, which aims to reduce world poverty by half (World Bank, 2002). Tourism, especially 'pro-poor' tourism, is seen as a way of addressing this MDG (see Saarinen, Chapter 15 in this book). Pro-poor tourism emphasizes utilization of natural resources, culture and heritage, which the poor have in abundance thereby developing sustainable livelihoods (Ashley *et al.*, 2000; Roe & Khanya, 2001; Rogerson, 2006).

Unlike other sectors where the product is brought to the consumer, the consumer of the tourism products (tourist) has to travel to the destination to be able to consume the product. This therefore enables linkages of tourism with other industries. This is an area that has received a lot of attention in the literature reviewed. Tourism provides markets for SMEs, especially markets run by women. Markets play a major role in empowering women. Traditional role prescription requires women to provide for the family. Markets in places such as Ghana, and many other developing countries, have become viable means of livelihood for women and have given women economic and political independence (Owusu & Lund, 2004).

Indeed, tourism does not only give women economic independence and empowerment, but it also fosters a sense of pride in local traditions and culture. This can be displayed in many forms, for example, through ethnic art, religious functions, producing handicrafts, using natural medicines, speaking local dialects, wearing traditional dress, and performing traditional songs and dances (Box 4.1). For example, claims that the production of ethnic art by Kuna women in Panama and Sani women in China serves as a viable way to resist cultural assimilation. Lama (1998) also agrees that women are the custodians of cultural values, traditions and knowledge because, during the colonial times and after, many men found employment in the cities, way from the villages. Women, on the other hand, were left in rural areas where they remained hardly influenced by Westernization and were therefore able to retain traditional beliefs and practices.

Box 4.1   Tourism, Gender and the Exotic 'Other': The Umhlanga Ceremony (Reed Dance) in Swaziland

JARKKO SAARINEN

The Umhlanga ceremony (Reed Dance) is a traditional dance where thousands of Swaziland's maidens perform the dance in their traditional attire. The ceremony lasts for about a week and takes place at the end of August or early September when the reed is matured and ready for harvest; however, the exact dates are set according to the moon's cycle. The ceremony presents the unmarried and childless maidens with an opportunity to honour the Queen Mother: the reeds cut and carried by the maidens are placed back at the main Royal household (Plate 4.1), where they are used as wind breakers for the Queen mothers hut. The other specific aim of the ceremony is to preserve girls' virginity by encouraging young Swazi women to abstain from sexual activities. This can be seen worthwhile in a country with a history of being a sex tourism destination originating from the Apartheid era of South Africa (Harrison, 1994) and especially under the present very high rate of HIV/AIDS-positive people: for the whole population the estimations vary between 18% and 39%, and according to UNAIDS (2007) the rate is highest among women (see UNDP, 2008; USAID, 2008). In addition, the ceremony with the related preparations aims to promote unity among the nation and girls by working together (see Swaziland Review, 2003).

Although the dates for the Reed Dance are not known in advance, which makes the advance marketing of the ceremony difficult, the dance may attract thousands of tourists to Swaziland. The performing Swazi women come from different parts of the country and their number can reach up to 20,000–40,000. The ceremony may also include a selection process of a new wife by the King of Swaziland (see Swaziland Review, 2003).

Although the Reed Dance can be seen as a good example of how local cultures celebrate their heritage, practice and share their traditional knowledge between generations and aim to promote acceptable behaviour (see Ivanovic, 2008: 155), there are also ethically problematic elements involved. As Ivanovic states in the context of the Zulu Reed Dance, the way in which young women are exposed to the eyes of tourists is questionable and recently also strongly criticized. The touristic commodification of the ceremony with related uses in the international promotion often creates images and representations rooted on the colonialism and the images and meanings of discoveries of the African continent by the Western male (van Eeden, 2006). According to Edwards (1996: 208), the creation of the 'exotic other' through exhibiting rituals and dance, for example, is still evident in tourism. In the case of the Umhlanga ceremony, the eroticized touristic landscape is dominated by females performing bodies referring to the colonial era and the primitive, exotic and erotic objects for Western (male) tourists to gaze at (Saarinen & Niskala, 2009), which may cause sexual harassment in the forms of encouragement for paid sex services and offered marriages (see Ivanovic, 2008: 155).

Indigenous knowledge systems have been promoted as sustainable tourism development. Women and girls have wide knowledge of the sustainable use of natural resources. Women more than men are dependent on natural resources products and crafts made from reeds and grasses. Maasai women, for example, have special knowledge of walking routes, craft production and useful plants, all of which can be related to the development of tourism products and services (Van der Cammen, 1997).

Globalization has resulted in the privatization and liberalization of businesses and tourism is no exception (Williams, 2002). Most developing

Plate 4.1 Maidens are carrying the reed to the Queen Mother at the Umhlanga (Reed Dance) ceremony in Swaziland (Photograph courtesy of Jarkko Saarinen, 2008)

countries are anxious to attract direct investment into their economies. Many popular sites, such as the Okavango Delta in Botswana, are operated by foreigner-owned companies (Mbaiwa, 2004; see Mbaiwa and Darkoh, Chapter 12 in this book). These companies have better facilities than the small- to medium-sized operations owned by women. As a result, SMEs are disadvantaged as they cannot compete with the well-resourced multi-national businesses.

Community-based Tourism

Some writers have suggested an approach to tourism that resonates with the Community-Based Natural Resources Management (CBNRM) philosophy. The assumption of CBNRM is that if communities are significantly involved in decision-making regarding the development, promotion and usage of tourism resources and can see the benefits thereof, they are likely to protect and conserve the natural resources on which tourism is based (Murphy, 1985).

In Belize, the Sandy Beach Women's cooperative runs a very successful ecotourism business specializing in eco-friendly accommodation and nature tourism. Involvement in the cooperative business has had positive lasting impacts on the Belize women. They have acquired business skills, marketing, environmental education and economic independence. Multiplier effects stem from the use of local materials and labour in the construction of lodges, preparation of local food, sales of crafts from lodges, and work for locals as cultural performers, demonstrators of traditional food cooking and as guides.

There have also been counter-arguments that tourism can perpetuate the inequalities that already exist in society because poor people do not have the resources to undertake tourism ventures. Secondly, they do not have the business skills that are important for running a tourism business. Consequently, most tourism ventures are controlled and owned by local elites or multinational companies who expropriate the capital thereby promoting leakages to developed countries where such corporations have their head offices (Jamal & Getz, 1995; Mbaiwa, 2004).

Tourism Economy and Women

Employment in the formal labour market

Tourism is also seen as creating employment opportunities for people and therefore contributing towards the development of their economic lives and the economy of the country in general. For example, the WTTC/WEFA research shows that in 2000, tourism generated directly and indirectly 11.7% of GDP and nearly 200 million jobs in the worldwide economy. WTTC/WEFA further suggests that a large proportion of women, minorities and young people were employed. As such, the UN regarded tourism as the world's largest industry and creator of jobs across national and regional economies.

In most developing countries there is high unemployment and because women have very low-level education and skills, they make up the majority of the unemployed. However, tourism is the largest employer of people with low-level skills and offers better opportunities for women, compared to other industries (Cattarinich, 2001). Therefore, the money earned through tourism has a multiplier effect on the economy, which in turn promotes development.

Contrary to the above view, it has been argued that organizational cultures are averse to women and use human resource policies to disadvantage women. Women are generally located in jobs stereotypically associated with their nurturing and care-giving roles in society. These are jobs at the lower levels of organizations with fewer opportunities for promotion to senior management (Manwa, 2005). As a result, women miss out on the opportunity to set the agenda and formulate the long-term strategic focus of the organization (Manwa, 2003; Parrett, 1999; Purcell, 1997; Skalpe, 2007).

Lower level jobs are generally casual, temporary or part time. These jobs offer less economic stability and make it difficult to balance domestic and productive work roles. Women cannot sustain their livelihoods from these jobs. They must instead seek alternative forms of livelihood since they cannot rely on tourism jobs to provide for their full upkeep. Women working in the tourism industry therefore rely heavily on financial contributions from other family members, for example, partners or children (Purcell, 1997; Skalpe, 2007; Wilson *et al.*, 2001).

Women's influence and decision-making around tourism

In several instances, women face the challenge of being taken as part of the product, with only attractive and young females securing employment in the industry as they are perceived to be able to satisfy the sexual needs of the tourists. Consequently, women are expected to dress in an 'attractive' manner and to look beautiful. Often the slim, young and pretty are preferred. They are then expected to 'play along' with sexual harassment by customers (Griffiths, 1999).

Women's social and economic empowerment

Tourism can also alleviate some of the social burdens that women often bear. For example, it can bring about the infrastructure that eliminates the need to travel long distances to fetch water for household use. Gurung (1995) (cited in Scheyvens, 2000) discusses how trekking tourism in the Annapurna area of Nepal had brought a number of benefits to the Dhampus village community. For example, when lodge owners installed water taps, this benefited many village women who otherwise had to walk long distances to fetch water. In addition, the demands of tourists for better facilities saw the adoption of labour-saving technologies – including kerosene stoves and solar water heaters that reduced the burden on women considerably.

Cases Studies from Southern Africa

The experiences of countries in southern Africa do not differ significantly from those states that have been mentioned in the literature reviewed above. More importantly, as Galvin (2007) has noted, SADC pays lip service to gender issues, hence the failure to agree on a gender policy for the region. The next section looks at tourism impacts on women in the formal labour market.

Employment in the formal labour market

Gendering of jobs is prevalent in the southern African region. In most southern African countries, the majority of tourism employees are women. Malawi and Zimbabwe are exceptions where the majority of employees are predominantly male (Manwa, 2002). These countries were part of the Federation of Rhodesia and Nyasaland (Malawi, Zambia and Zimbabwe) where the colonial government employed men even in domestic jobs generally associated with females. In the SADC region, there is prevalent gendering of jobs whereby women dominate the middle to lower levels of organizations (Manwa, 2003). The RETOSA board as shown in Table 4.1 typifies the prevailing scenario in most southern African countries. RETOSA's board is made up of senior managers in tourism ministries (permanent/deputy permanent secretary) and CEOs of private sector organizations. As evident from Table 4.1, strategic decision-making positions in the tourism industry are still a preserve of men.

The private sector seems to offer better opportunities for women to advance to the executive suites. Table 4.1 shows that, of the six women who were members of RETOSA's board, four were from the private sector. The rationale could be that such organizations are generally perceived to perform public relations exercise, which is generally seen as a female job (Manwa, 2005). In contrast, the majority of Chief Executive Officers of public enterprises popularly known as Tourism Boards, or Authorities, are men (RETOSA, 2007). In a few instances where CEOs were women, the majority of the board members and senior managers were men. These results confirm the trend observed elsewhere concerning the paucity of women in executive suites of both the private- and public-sector organizations.

Studies in Zimbabwe and Lesotho have also confirmed that women still opt for jobs with fewer chances for promotion (Manwa, 2003). For women to make it into the executive suites of corporations is still a rarity in the southern African region. When they do, their achievements get a lot of publicity in the media. For example, the Hotel and Catering Magazine of October/November 2007 featured the first woman Managing Director

Table 4.1   RETOSA Board (2005–2006) by gender and sector

	Male		*Female*	
Sector	*Number*	*%*	*Number*	*%*
Private	4	20	4	20
Public	10	50	2	10
Total	14	70	6	30

Source: RETOSA Strategic and Business Plan for Regional Tourism in SADC 2003–2007

of Shearwater Tourism Group in Zimbabwe, as well as a few other women who have made it to the top in the Botswana tourism industry, such as the Chief Executive Officer of the Botswana Tourism Board.

Empowerment and influence and decision-making

The CBNRM model is supposed to be a grassroots initiative. But grassroots approaches as noted by the IUCN Botswana (2007) do not take cognizance of inequalities women face even at grassroot levels. CBNRM currently focuses on wildlife. This is a resource owned by men in communities. Women, on the other hand, use natural products such as firewood and wild grasses, on which the livelihoods of rural communities depend. The predominant patriarchal culture relegates women to the status of minors who only implement decisions made by men (Manwa, 2002; Thabane, 2000). Examples include fencing off of water wells and firewood in an effort to protect villagers' fields from wild animals in Zimbabwe (Nabane, 1994). If women were involved in such decisions they would have suggested the creation of alternative sources of water and fire wood, such as digging of boreholes in proximity to the homesteads as well as finding alternative sources of firewood (Manwa, 2003).

Another good example is the Lesotho Highland Water Project (LHWP) (IUCN, 1998), which was established by a treaty between the Governments of Lesotho and South Africa in 1986. The project started without any environmental impact assessment being undertaken. It involved moving people to either some parts of the highlands or resettling them in the lowland areas. The project aims at increasing tourism arrivals (LHWP), among other benefits to the economy of Lesotho (Lang, 2008).

The project has had serious negative impacts on people living along the Malibamatso River in the Highlands of Lesotho. Less water availability downstream of the scheme means that there is now less agricultural output in some villages. Similarly, the flooding of some villages' agricultural fields has caused reduced output. Firewood, which used to be accessible along the river banks, has also been made inaccessible as a result of the flooding of the dam. This means that women now have to travel long distances in search of firewood and water. There are also other livelihoods losses to the communities. The IUCN Botswana (2007) shows resources used by the communities who have been moved from the areas:

(1) Fish caught by 5098 households, which caught an annual average of 17.5 kg of smallmouth yellowfish, 2.2 kg of rock catfish and 3.0 kg of rainbow trout. The approximate market value is US$1.4 per kg.

(2) Wild vegetables harvested by 13,911 households, which gathered on average 148 plastic bags per year. The mean market value is US$0.3 per bag.

(3) Reeds, thatch grasses and the craft grass *Leloli* within the riparian zone are harvested by 6713, 7972 and 5487 households, respectively.

(4) Medicinal plants within the riparian zone are collected by 6391 households, with a mean market value of approximately US$6.6 annually per household.

(5) Construction of the Katse Dam has resulted in the decrease and extinction of food resources for smallmouth yellowfish, loss of waterbirds and medicinal plants.

The above data demonstrate adverse economic impacts on women, who are the main users of the resources affected.

Local women are often overlooked when lodges and other ecotourism sites are developed. In the Mahenye district of Chipinge in Zimbabwe, for example, a joint venture agreement between the local Shangaan people and Zimbabwe Sun Limited, which owns a chain of hotels in the country, has seen the development of two tourist lodges on Shangaan land. Up until the early 1990s, the Shangaan had poached extensively from the neighbouring Gonarezhou National Park, from which many of them were evicted and their villages burnt down when the park was created in 1966. The agreement with the Zimbabwe Sun Ltd has brought a lot of development to the area. Employment at the lodges has been heavily biased towards men, with the result that only 4 out of 48 lodge workers are women (Scheyvens, 2005).

Some of the positive impacts of CBNRM include recognition and revitalization of local cultures. The obvious manifestations of culture include production of traditional crafts and traditional singing and dancing. Such activities are seen as building a sense of pride and self-worth, as well as a means of preserving cultural identities. Since most of the people involved in these activities are women, it would seem that they are benefiting more. The basket making in the Okavango Delta in Botswana (Mbaiwa, 2004), and the Vulamehlo Handicraft project in Kwazulu Natal in South Africa (Kruger & Vester, 2001) are examples of successful projects run by women under CBNRM.

The production of baskets and beads from ostrich egg shells in Botswana's Okavango is of economic value to Botswana women as this is the skill that is acquired mostly by women and has become one of the important sources of livelihoods in the area. In the 1980s, it provided self-employment to between 400 and 1500 women (Terry, 1994). Money acquired from selling the baskets is used for buying food, school uniforms for children and clothing for the family.

The Vulamehlo Handicraft is aimed at uplifting the living standards of women in KwaZulu Natal. Women are able to work from home and thereby lessen disruption of family life while still undertaking their traditional family roles. Polygamy is common in Zulu culture and women are often abandoned by their partners, temporarily or permanently. Thus, economic independence is a matter of survival for many of them (Scheyvens, 1999). The cooperative has increased the self-reliance and pride of its members by providing them with a reliable source of income so they can, among other things, afford the fees to send their children to school. Some of the life-changing benefits of the project have been the women's continued production of skilled crafts and the learning of business management skills while they are earning an income for the family. Although it is true that the income from such activities may not be able to support all the family's needs, nevertheless it is an important supplementary source to family income. Other benefits include environmental protection, preservation of culture and skills.

Case studies have shown that the participation of women in tourism has empowered them by giving them economic independence, self-confidence and resources needed for starting their own business. This also reduces reliance on men as well as the pressure to get married at a young age (McKenzie, 2007).

Conclusion

The case studies presented in this chapter show the varying degree to which women are involved in tourism, and also the nature of their involvement. What is clear though is that women are mainly on the peripheries of tourism development. I would therefore like to conclude this chapter by making the following recommendations that could somehow help improve the position of women in tourism in southern Africa (see Figure 4.1).

Figure 4.1 highlights linkages and interrelationships of intervention strategies that could be adopted. Firstly, critical to any intervention strategy is the acknowledgement of the centrality of societal culture, which as Kinnaird and Hall (1996) have noted, influences people's behaviour in all spheres of their existence. Culture can be used as a vehicle for change in women's involvement in tourism development in the SADC region.

To conclude, SADC governments should come up with legal instruments to improve the position of women in tourism, and SADC governments should develop monitoring and evaluation instruments to assess the progress of women in the industry, and to mitigate any negative outcomes, while learning from best practices internationally. SADC governments should also recognize the important contribution of tourism SMEs and the role of women in them, by providing funding and training, and if the government is proud of the national culture and heritage the population is bound to follow suit. Therefore they must come up with motivation to instil pride in existing local competencies, skills and culture. There should be a concerted effort to develop cultural and heritage tourism in order to

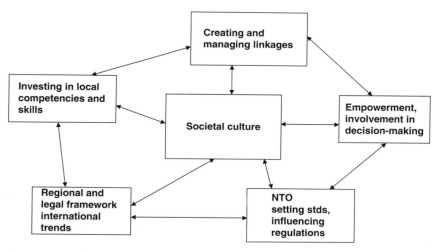

Figure 4.1 Inter-related factors to empower women

empower women, who are custodians of culture. Above all, whatever decisions are made regarding SMEs should be arrived at in consultation with the affected women.

The national tourism organizations/authorities can play a major role in market analysis (Moscardo, 2005). Most SADC destinations market mainly the key attractor to a destination. For example, from my experience and observation, the Victoria Falls in Zimbabwe is marketed as The falls with only the major attractions mentioned. These include watching the falls, game drives, helicopter flyover, skydiving, bungee jumping and white water rafting. However, in the same locality one also finds SMEs selling crafts, various dance groups representing the San, Tonga, Ndebele and other ethnic groups, as well as traditional foods that are not marketed and do not feature in both print and electronic media advertisements about Victoria Falls. No effort has been taken to profile SME operators, the business they are in and their special needs. In most cases, these peripheral operators, who are mostly women (Moscardo, 2005), are reduced to hawking on the tourist trails and are consistently harassed and even prosecuted for harassing tourists or accused of robbing them. National Tourism Boards could play a key role in marketing these SMEs.

As Moscardo (2005) has noted, the women-dominated peripheral tourism, has failed to grow because of various factors, including the women' inability to analyse the market and the potential customers and failure to understand the product offering. This is where tourism boards can come in.

References

Ashley, C., Boyd, C. and Goodwin, H. (2000) Pro-poor tourism; putting poverty at the heart of the tourism agenda. *ODI Natural Resource Perspectives* 51, 1–6.

Brohman, J. (1996) New directions in tourism for third world development. *Annals of Tourism Research* 23 (1), 48–70.

Cattarinich, X. (2001) Pro-poor tourism initiatives in developing countries: Analysis of secondary case studies. On WWW at http://www.propoortourism.org.uk/initiatives_cs.pdf. Accessed 9.8.2007.

Dao, M.Q. (2004) Rural poverty in developing countries: An empirical analysis. *Journal of Economic Studies* 31 (6), 500–508.

Edwards, E. (1996) Postcards: Greetings from another world. In T. Selwyn (ed.) *The Tourist Image. Myths and Myth Making in Tourism* (pp. 197–222). Chichester: Wiley.

van Eeden, J. (2006) Land rover and colonial-style adventure. *International Feminist Journal of Politics* 8 (3), 343–369.

Galvin, T. (2007) Southern African gender policy – where gender and development meet. Paper presented at SAUSSC conference held in Manzini, Swaziland, 25–29 November.

Ghimire, K.B. and Pimbert, M.P. (1997) Social change and conservation: An overview of issues and concepts. In K.B. Ghimire and M.P. Pimbert (ed.) *Social Change and Conservation* (pp. 297–330). New York: Earthscan.

Goodwin, H. (2000) Pro-poor tourism: Opportunities for sustainable local development. *Development and Cooperation*, 5 (September/October), 12–14.

Griffiths, B. (1999) Women's sexual objectification in the tourism industry in the United Kingdom. In M. Hemmati (ed.) *Gender and Tourism: Women's Employment and Participation* (pp. 35–47). London: UNED-UK.

Gurung, C.P. (1995) People and their participation: New approaches to resolving conflicts and promoting cooperation. In J.A. McNeely (ed.) *Expanding Partnerships in Conservation* (pp. 223–233). Washington DC: Island Press.

Hall, D. (1999) Understanding tourism processes: A gender-aware framework. *Tourism Management* 17 (2), 95–102.

Harrison, D. (1994) Tourism and prostitution: Sleeping with the enemy? Case of Swaziland. *Tourism Management* 15 (6), 435–443.

IUCN (1998) Lesotho Highlands Water Project. On WWW at http://www.lhwp.org.ls/default.htm. Accessed 12.12.2008.

IUCN Botswana (2007) Botswana CBRNM support programme: The role of women in CBNRM (IUCN Report). On WWW at http://www.cbnrm.bw/cbnrmbotswana/gender.html. Accessed 12.11.2007.

Ivanovic, M. (2008) *Cultural Tourism*. Cape Town: Juta.

Jamal, T.B. and Getz, D. (1995) Collaboration theory and community tourism planning. *Annals of Tourism Research* 22 (1), 186–204.

Kinnaird, V. and Hall, D. (1996) Understanding tourism processes: A gender-aware framework. *Tourism Management* 17 (2), 95–102.

Kruger, S. and Vester, R. (2001) An appraisal of the Velamehlo Handicraft Project. *Development Southern Africa* 18 (2), 239–252.

Lama, W.B. (1998) CBMT: Women in the Himalaya. Presented to the Community-based Mountain Tourism conference posted on the Mountain Forum Discussion Archives. On WWW at: http://www.mtnforum.org/mtnforum/archives/document/discuss98/cbmt/cbmt4/050898d.htm. Accessed 10.4.2008.

Lang, S. (2008) Development: Unexpected benefits of Lesotho Highlands Water project. On WWW at http://ipsnews.asp?idnews=41005. Accessed 28.3.2008.

Manwa, H. (2002) 'Think Manager, Think Male' does it apply to Zimbabwe? *Zambezia* 29 (1), 60–75.

Manwa, H. (2003) Wildlife-based tourism, ecology, sustainability: A tug-of-war among competing interests in Zimbabwe. *Journal of Tourism Studies* 14 (2), 45–54.

Manwa, H. (2005) Gender issues impacting on the delivery of services in the tourism and hospitality industry. Paper presented at the Hospitality and Tourism Association of Botswana Annual Briefing.

Mbaiwa, J.E. (2004) Prospects of basket production in promoting sustainable rural livelihoods in the Okavango Delta, Botswana. *International Journal of Tourism Research* 6, 221–235.

Mckenzie, K. (2007) Belizean woman and tourism work; opportunity or impediment. *Annals of Tourism Research* 34 (2), 447–496.

Moscardo, G. (2005) Peripheral tourism development: Challenges, issues and success factors. *Tourism Recreation Research* 30 (1), 27–43.

Mowforth, M. and Munt, I. (1998) *Tourism Sustainability. New Tourism in the Third World*. London: Routledge.

Murphy, P. (1985) *Tourism. A Community Approach*. London: Methuen.

Nabane, N. (1994) A gender sensitive analysis of a community based wildlife utilisation initiative in Zimbabwe's Zambezi valley. CASS Occasional paper series, NRM.

Owusu, G. and Lund, R. (2004) Markets and women's trade: Exploring their role in district development in Ghana. *Norwegian Journal of Geography* 58 (3), 113–124.

Parrett, L. (1999) London Thames Gateway Forum. Women in Tourism Employment. A Guided Tour of the Greenwich Experience.

Purcell, K. (1997) Women's employment in UK tourism. In M.T. Sinclair (ed.) *Gender, Work and Tourism* (pp. 33–59). New York and London: Routledge.

RETOSA (1997) *RETOSA Charter*. Midrand: SADC.

RETOSA (2007) *Strategic and Business Plan for Regional Tourism in SADC 2007/8–2011/12*. Midrand: RETOSA.

RISDP (2002) SADC Regional Indicative Strategic Development Plan. On WWW at http://www.sadc.int. Accessed 12.12.2008.

Roe, D. and Khanya, P.U. (2001) *Pro-Poor Tourism: Harnessing the World's Largest Industry for the World's Poor*. London: International Institute for Environment and Development (IIED).

Rogerson, C.M. (2006) Pro-poor local economic development in South Africa: The role of pro-poor tourism. *Local Environment* 11 (1), 37–60.

Saarinen, J. and Niskala, M. (2009) Selling places, constructing local cultures in tourism – the role of Ovahimbas in Namibian tourism promotion. In P. Hottola (ed.) *Tourism Strategies and Local Responses in Southern Africa*. Wallingford: CABI Publishing (forthcoming).

SADC (1992) *Southern African Development Community Treaty*. On WWW at http://www.africa-union.org/root/au/recs/sadc.htm#commun.Accessed 9.3.2008).

SADC (1998) *Protocol on Tourism Development*. Gaborone: Botswana.

SADC (2002) *Regional Indicative Strategic Development Plan*. On WWW at http://www.sadc.int/content/english/key.documents/risdp. Accessed 29.2.2008.

Scheyvens, R. (1999) Ecotourism and the empowerment of local communities. *Tourism Management* 20 (2), 245–249.

Scheyvens, R. (2000) Promoting women's empowerment through involvement in ecotourism: Experiences from the Third World. *Journal of Sustainable Tourism* 8 (3), 232–248.

Scheyvens, R. (2005) Growth of beach tourism in Samoa: The high-value low cost? In C.M. Hall and S. Boyd (ed.) *Nature-based Tourism in Peripheral Areas. Development or Disaster?* (pp. 188–202). London: Channelview Publications.

Skalpe, O. (2007) The CEO gender pay gap in the tourism industry – evidence from Norway. *Tourism Management* 28 (3), 845–853.

Swaziland Review (2003) *Tourism*. Mbabane: Chamber of Commerce.

Terry, M.E. (1994) The Botswana handicraft industry: Moving from 20th to 21st century. In J. Hemos and K. Ntetas (ed.) *Botswana in the 21st Century* (pp. 571–583). Gaborone: Botswana Society.

Thabane, M. (2000) Shifts from old to new social and ecological environments in the Lesotho Highlands Water Scheme; relocating residents of the Mohale Dam area. *Journal of Southern African Studies* 26 (4), 633–654.

UNAIDS (2007) *AIDS Epidemic Update 2007*. On WWW at http://data.unaids.org/pub/EPISlides/2007/2007_epiupdate_en.pdf. Accessed 12.8.2008.

UNDP (2008) Responding to HIV & AIDS in Swaziland. On WWW at http://www.undp.org.sz/hivaids.htm. Accessed 9.8.2008.

USAID (2008) Health profile: Swaziland. On WWW at http://www.usaid.gov/our_work/global_health/aids/Countries/africa/swaziland_05.pdf. Accessed 9.8.2008.

Van der Cammen, S. (1997) Involving Maasai women in tourism. In L. France (ed.) *The Earthscan Reader in Sustainable Tourism* (pp. 161–163). London: Earthscan.

Williams, M. (2002) Economic literacy series: General Agreement on Trade in Services Number 5: Tourism and liberalisation, gender and the GATS, International Gender and Trade Network, Secretariat.

Wilson, S., Fesenmaier, D.R., Fesenmaier, J. and Van es, J.G. (2001) Factors for success in rural tourism development. *Journal of Travel Research* 40 (2), 132–138.

World Bank (2002) *Stratégie de développement touristique en Tunisie: Rapport phase I*. Washington DC: World Bank.

Chapter 5

Sustainable Tourism: Perspectives to Sustainability in Tourism

JARKKO SAARINEN

Introduction

To use an old phrase, uncontrolled tourism can saw its own branch to sit off if it only aims at growth in quantity – in tourist numbers, facilities and total turnover. It is therefore important to the long-term benefits and survival of the tourism industry itself to develop a better understanding of its impacts and its limits for growth (Butler, 1999). Therefore, the idea of sustainability has become an important policy issue in tourism management and development. For example, the United Nations World Tourism Organisation (UNWTO), among many other international institutions, has strongly emphasised a need for sustainability and recognition of the limits of growth in tourism (see WTO, 1993).

Similarly, southern African countries and their regional development organisations are committed to the idea of sustainability in tourism policies (see Askeeke & Katjiuongua, 2007; Government of Botswana, 2002; Department of Environmental Affairs and Tourism, 2002). The background of these sustainable tourism development policy needs has been the growing role of tourism in the region and evolved concerns over the local level benefits of the increasingly global industry; tourism is seen as a vital economy but which should benefit the local communities and residents of the places that tourists actually visit. By this emphasis, governments and regional development agencies see tourism as a process that can lead to well-being and higher living standards by encouraging people to value and conserve the physical and cultural environment.

Nowadays sustainability is linked to almost all kinds and scales of tourism activities, including mass tourism (Clarke, 1997). However, there are still disagreements about the practices and meanings of sustainability and its usability in tourism (Bianchi, 2004; Butler, 1991; Liu, 2003; Sharpley, 2000). Some scholars even suggest that the sustainability in tourism policies is more rhetoric and 'greenwashing' instead of representing a serious attempt towards sustainable development (Butler, 1999; Mowforth & Munt, 1998; Wheeller, 1993). This chapter aims to introduce and critically discuss the idea of sustainability in tourism development discussions. The purpose is to provide perspectives serving the following chapters focusing on the specific issues and cases in southern Africa. This is done by evaluating the different ways to understand the sustainability and recognising that behind the distinct perspectives there are traditions that are different in their focuses and goals. These are referred to as resource-, activity-, and community-based traditions of sustainability in tourism (Holden, 2007; Saarinen, 2006), which are characterised by different ideas of the general nature and character of the limits of growth and how they can be known and defined. Finally, the conditions under which (sustainable) tourism could represent a tool for wider sustainable development are discussed in the conclusions.

Traditions of Sustainability in Tourism

The resource-based approach

The earliest discussions on the limits of growth in tourism were related to the carrying capacity model and a search for the magical number which cannot be overstepped without serious negative impacts on the resources available (Wagar, 1964). In the tourism and recreation context, the concept of carrying capacity has been defined as the maximum number of people who can use a site without any unacceptable alteration in the physical environment and without any unacceptable decline in the quality of the experience gained by tourists (Mathieson & Wall, 1982: 21). This approach towards the limits to growth in tourism development refers to a resource-based tradition in research. Historically, it is especially related to recreation studies in natural or semi-natural settings (Lucas, 1964; Stankey, 1982), the basic assumptions being derived from late 19th century livestock and wildlife management studies (Pigram & Jenkins, 1999: 90). The roots of the research tradition are deeply founded in positivism and the natural sciences (see Buckley, 1999).

The resource-based approach implies an objective and measurable limit or stage of growth at which there is no room for any more tourists and tourism activities in a certain environment. To achieve further growth and development in tourism, people and the industry will have to cope with the environment in a new and better way, such as by altering their activities or number but not primarily the resource that is used. In recreation and tourism studies this has led to density, erosion, disturbance, crowding, social carrying capacity and authenticity analyses, for example (Anderson & Brown, 1984; Aronsson, 1994; Vaske *et al.*, 1986).

In the resource-based tradition, the limits to the growth and impacts are evaluated in relation to the resources used in tourism and the assumed or known natural or original (non-tourism) conditions (Buckley, 2003). The limit of growth – carrying capacity – is regarded as objective and measurable in nature in terms of the original characteristics of the resource, that is, what the space used for tourism would be without tourism-related activities. The indicators of resource-based sustainability reflect the relation between the condition of the resources and the impacts of development. Therefore, the main subject of evaluation is the resource (physical and/or social environment) and its condition, and the limits of growth are defined by comparing the condition of the resources used with that of a similar environment not used in activities, or by describing and evaluating the intensity of the physical, social or cultural changes resulting from tourism (Hammitt & Cole, 1987; Taylor, 2001).

The challenges of the resource-based approach are how to define the original non-tourism conditions of the resources and separate the impacts of the industry from changes caused by other activities and natural or human-induced processes at destinations (Collins, 1998). In addition, tourism always causes some impacts, which leads to the critical question of which impacts are objectively acceptable and to what degree; this is a question of human values and perceptions concerning the resource, indicators, criteria and impacts. Thus, the search for a magical absolute and objective calculation of the maximum number of tourists at a destination has failed in most cases (see Butler, 1996; Hughes & Furley, 1996; Lindberg *et al.*, 1997). For example, many of the popular national parks in southern Africa, such as Kruger (KNP) and Pilanesberg (PNP) National Parks in South Africa adopted calculative limitations to visitor numbers. According to Ferreira and Harmse (1999), the late 1990s management plan in the KNP set the maximum number of tourists at 10,552 visitors per day (overnight visitors at 6902 and day visitors at 3650). However, as they correctly state, a definitive formula to set this kind of limit for the touristic use of the park was not realistic. Recently, many of the national parks in southern Africa and elsewhere have turned to management policies without having to set rigid limitations on entry levels (see Kruger National Park Management Plan, 2006; Pilanesberg Management Plan, 2006; see Mabunda & Wilson, Chapter 7, this volume), which also allows a wider spectrum and scale of tourist activity developments.

The activity-based approach

Tourism is a dynamic activity and it is for this reason that the resource-based idea appears to be problematic for the industry and its development models. Tourism and its changes produce impacts and, because of the broad range of interests, some are perceived as negative. McKercher (1993) has aimed to overcome the problem by emphasising the needs and active role of the industry in defining sustainability and the limits to growth.

Development and industry-oriented solutions for sustainable tourism can be assigned to an activity-based tradition of sustainability, implying that certain tourist activities, or the industry itself, may have a maximum capacity (Wall, 1982).

Unlike the case of resource-based sustainability, individuals and human activities in tourism do not necessarily alter their behaviour in the first instance in their relation to the resources used in tourism. To grow and develop, the industry and other related actors can, and will, modify the environment, the resources, for their needs. In a national park context this is done by establishing a zoning system that allows different kinds of activities and intensive tourism developments in some areas while leaving some parts of the park unused by tourism. KNP's zonation plan, for example, involves five types of areas ranging from a high-intensity leisure zone to a wilderness zone (Kruger National Park Management Plan, 2006). The former category allows park management and business to establish an intensive tourism infrastructure and a multitude of tourist activities, whereas the latter restricts the touristic use and entry to such areas inside the park. Between the two ends there are three mid-categories termed as low-intensity leisure, primitive and remote, which offer diverse levels of opportunities for tourism and recreation activities.

Obviously, the activity-based tradition is more industrially oriented than the resource-based tradition. It refers to tourism-centric approaches in development discussions (see Burns, 1999), focusing more on the needs of tourism as an economic activity. In tourism research it relates to the tourism area cycle of evolution, for example, which was derived from the product lifecycle idea in marketing studies and combined with the idea of limited resources and carrying capacity (Day, 1981; Stansfield, 1978). This well-known tourism lifecycle model proposed by Richard Butler (1980) describes the change process of a destination from the early exploration and involvement stages through the development and consolidation stages, and finally to the stagnation stage that is alternatively followed by new growth or decline (Figure 5.1).

According to Butler, every tourism area has a limit to its growth and the stagnated situation implies that this limit has been reached. The element that limits growth, referred to in the model as carrying capacity, is depicted as a zone that controls the scale of development in a specific environment. This interrelationship between tourism development and carrying capacity, however, is a dynamic one (Butler, 1997: 116; Martin & Uysal, 1990). Although unlimited growth at any destination is impossible, development may be cyclical in nature (Baum, 1998; Butler, 2004): as previously mentioned after the stagnation stage of the model, or even before if new major products or marketing schemes have been introduced, the development cycle can begin again and exhibit new (absolute) growth (Tooman, 1997). These potential multiple and constantly rising

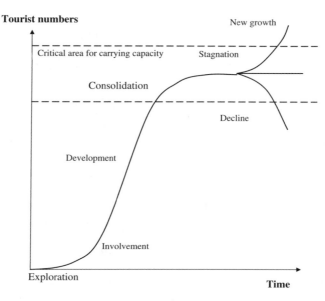

Figure 5.1 Tourism area life cycle model (after Butler, 1980)

cycles in the evolution model may challenge the role of carrying capacity and its connection with the resources used in tourism during the changing cycles. Thus, the limit of growth in the evolution model is not primarily based on the capacity of the destination and its ('original') resources for absorbing tourism as a small-scale nature-based tourism lodge or local picturesque fishing village, for example, but on the tourism as activity and its capacity to change the environment (see Getz, 1983). By changing the tourism product (destination) through development and marketing, and by introducing new types of activities, facilities, infrastructure and so on, the destination and its limits of growth can be modified and moved forward to a new level.

Indeed, many of the old mass tourism destinations have started as small-scale attractions for regional and national demand but after several cycles of development they have turned into international scale tourist complexes. However, all touristic modifications based on the development of new cycles will potentially require more effective and massive environmental changes, new land-use patterns and additional construction work, all of which can easily overstep the limits of resource-based (such as ecologically, socially and culturally defined) sustainability. Therefore, the activity-based tradition involves a relativist approach. It implies that certain tourism activities have or may have different kinds of limits on their growth, or that certain industry segments have different abilities to cope with impacts and other tourists. Instead of the original resource utilised (as in the exploration, involvement and early development stages), the limits of growth are mainly based on specific changing activities, capacities or products. Thus, the subject of evaluation is tourism, tourist activities and their capacities for growth.

In this respect, the non-growth situation implies that the limit, in terms of carrying capacity and sustainability, may be reached and modifications are needed in tourism activities and products for further development. As such, the indicators of activity-based sustainability reflect the relation between tourism activities and development intensity of the industry. Activity-based sustainability is based on the idea of a dynamic, transforming tourism space, the limits of whose growth are evaluated based on the activities and their shifting needs and capacities for utilising the environment and other resources in tourism. From this perspective, the rigid limitations on areas such as national parks and their entry levels, for example, are problematic. The changed management orientation of national parks in South Africa, for example, is a result of these challenges in setting objective limits, and currently these changes reflect not only the new developmental role of conservation areas in regional economies but also the policy level emphasised needs of local communities to benefit and have some level of control over tourism development using their present or past living environments.

The community-based approach

The relation between resource-based and activity-based sustainability appears to be problematic. As tourism grows, indicating that the limits of activity-based sustainability have not yet been reached, development actions may, and often do, exceed the resource-based capacity to change. Efforts have been made in the literature and management procedures to overcome this dual nature of sustainability by invoking different negotiation and participation processes. These participation processes refer to community-based tourism approaches (Kieti & Akama, 2005; Murphy, 1983, 1988; Timothy & White, 1999). The setting of limits for growth through negotiation and participation places the hosts and the benefits that they may gain from tourism in a central position in the process (Robinson & Hall, 2000; Scheyvens, 1999).

Recently, participatory approaches have evolved towards new kinds of alternative processes, which include community-based and pro-poor tourism, all aiming at practices that contribute to the local bases and especially to the needs of marginalised people (Duffy, 2002: 100–102; Hall, 1994: 43–45). According to Jost Krippendorf (1982), community-based tourism as an alternative form of tourism development aims to ensure that tourism policies should no longer concentrate on economic and industrial necessities alone. Thus, there is a strong emphasis on the needs of local people, their physical and societal environment and wider needs in community-based resource management processes and policies (see Cater, 1993; Craik, 1995; Mowforth & Munt, 1998), which are strongly recognised in many national tourism policies in southern Africa (see Ministry of Environment and Tourism, 2005a).

In general, community-based tourism aims to ensure that members of the local communities have a high degree of control or even ownership over the tourism activities, its limits and resources used (Saarinen, 2006; Scheyvens, 2002). Local people should receive a significant share of the economic benefits of tourism in the form of direct revenues and employment, upgraded infrastructures, environment and housing standards and so on (Stronza, 2007) (Box 5.1). According to Telfer and Shapley (2007), there are two major goals in community-based tourism. First, it should be socially sustainable, which refers to the already-mentioned role of local control and participation in tourism operations and the shared socioeconomic benefits. Second, community-based tourism has respect for local cultures, identities, traditions and heritage.

Box 5.1   The Nata Sanctuary Community-based Project

The Nata Sanctuary Community-based Project is located on the tip of Makgadikgadi Salt Pans in north-eastern Botswana. The Sanctuary is part of the Makgadikgadi Pans Important Bird Area (IBA). The area provides opportunities to see Lesser Flamingos, Greater Flamingos, Great White and Pink-backed Pelicans, for example. To protect the area and its abundant bird life, a community-based project, with representatives from the surrounding communities of Manxotai, Maposa, Nata and Sepako, was established in the late 1980s (Plate 5.1). The project provides tourism services and products such as camping grounds, bird-watching and local crafts.

The main visitor segments are based on self-driving South African tourists and local school groups. Surplus revenue from tourism is used for capital development and environmental education within the four villages. At the beginning of the project, the surplus cash was given directly to the communities but lately the revenue has been mainly used to develop the sanctuary's infrastructure and facilities and the main benefits of the community-based tourism activities have been related to local employment.

Plate 5.1   The entrance of the Nata Sanctuary, Makgadikgadi Salt Pans. In October 2008, the main parts of the Sanctuary along the nearby private Nata Lodge were burned by extensive bush fires (Photograph courtesy of Jarkko Saarinen, 2008)

In Namibia, for example, community-based tourism is widely recognised in the national tourism development policies. The Ministry of Environment and Tourism (2005b) has initiated the Community-based Tourism Policy that aims to explore ways in which communities can benefit from the tourism industry that promotes social and economic development and conservation in communal areas (see Becker, Chapter 6 and Scholz, Chapter 9 in this book). The government also supports strongly the national community-based tourism association (NACOBTA, i.e. Namibia Community-based Tourism Assistance Trust), which was initiated in 1995 by local communities. As a non-profit organisation it aims to improve the living standards of people by developing sustainable community-based tourism and especially joint ventures integrating private sector investors with local communities (Massyn, 2007).

According to the Ministry of Environment and Tourism (2005b), the community-based tourism policy in Namibia involves the following benefits and conditions:

- People can benefit from tourism and participate in tourism planning.
- People can benefit from tourism on their land and conserve wildlife and natural resources.

- People will be encouraged to develop tourism enterprises.
- Development on communal land must be acceptable to the people living there.
- Established tourism businesses are encouraged to work with people in communal areas.
- Tourism development will work hand-in-hand with conservation of the environment.

The key issues in the policy – and community-based tourism in general – are the questions of participation and empowerment: how to integrate local communities in tourism planning and how to ensure a sufficient level of power and control in decision-making processes over the use of natural and cultural resources in tourism development (Saarinen & Niskala, 2009; Ramutsidela, Chapter 10 in this book).

In this respect the community-based approach implies that sustainability is, or can be, defined through a negotiation process, which indicates that the limits of growth are socially constructed (Bryant & Wilson, 1998; Hughes, 1995; Redcliffe & Woodgate, 1997). As a social construct, sustainability refers to the maximum levels of the known or perceived impacts of tourism that are permissible in a certain time–space context before the negative impacts are considered to be too disturbing from the perspectives of specific social, cultural, political or economic actors who possess sufficient power over the chosen indicators and criteria. The community-based tradition aims to empower the hosts in development discourses and practices, but at the end the constructive perspective indicates that the limits of tourism are associated with power relations in a certain context. By empowering the communities, however, the limits of growth in tourism can be defined in a more equable way and one that is more beneficial for the local people (Scheyvens, 2002).

From the community-based view, sustainable tourism and the limits of growth are understood as dynamic and contested ideas that are continually being constructed and reconstructed during the process of development and negotiations. The conceptualisation of tourism spaces and their sustainability as social constructs does not necessarily undervalue resource-based capacity and the realm of nature or ecological changes and their character in any objective or measurable sense. However, the community-based tradition indicates that the concept of sustainable tourism is not objective but social and related to knowledge. Thus, it is also laden with power issues. For example, such questions as who can define and decide what is an ecologically acceptable change, for example, in southern African national parks, or what resources should be sustained and for whom in these areas and adjacent lands, are all loaded with power issues. In most cases, the answers to these issues are not derived directly from the impacts themselves but from the historically contingent social, economic and political practices and discourses of the current power relations defining them. This is evident in the land reform and tourism and rural development discussions in southern Africa (see Fraser, 2007).

Conclusions

The idea of sustainable tourism has both fascinated and irritated academics and developers, and the concept has aroused harsh criticism from the beginning of its existence (see Butler, 1991; McKercher, 1993). One of the key problems is tied to the holistic nature of sustainability, especially its spatial and temporal scales. Tourism is a broad system based on the movement of people, goods, capital and ideas, among many other things, between home regions and destinations that are linked by means of routes and transit regions and associated with many other societal processes. Tourism is also increasingly becoming a part of the global economy and culture, but the focus of sustainability has nevertheless been mainly on destinations and tourism practices in those areas, grasping the most visible processes and impacts related to the industry, but only the fragment of the total.

This limitation of sustainable tourism is not only technical in nature but also ethical (Holden, 2003; Macbeth, 2005). In sustainable development, the issues of scale and the global–local nexus play an important role (Duffy, 2002), but as already indicated tourism has focused on contributing to sustainable development mainly on a local scale. Although this can be highly beneficial, this site, destination or community-specific approach is currently debated and challenged by global environmental ethics, which questions not only the limited scale but also the anthropocentric position, which is built on the idea of sustainability, by emphasising the rights of nature as equal to those of humans (Holden, 2003). Therefore, Holden (2007) has argued that the present challenge with sustainable tourism is to see it as more than just an industry and a destination scale management issue. According to him this leads to wider ethical considerations including tourism and its responsibilities in the context of global climate change, for example (see Gössling & Hall, 2005). Although tourism is greatly dependent on and influenced by changing climate, it is also an actor with responsibilities in that process of change.

Indeed, sustainable development is still a major challenge for the industry and tourism studies. Tourism has impacts on all geographical scales and therefore it also has some level of responsibility in different scales if it aims towards true sustainability (Holden, 2003; Saarinen, 2006). However, presently the three traditions discussed here represent the main aspects and elements of the idea of sustainability that are all referring to local-scale issues and challenges. The resource-based tradition reflects the limits of the original conditions of local resources and the needs to protect nature (natural capital) and the local culture (cultural capital) from unacceptable changes caused by tourism activities. In contrast to that the activity-based tradition refers to the resource needs of the industry with respect to its present and future development, aiming to sustain the economic capital invested in tourism. The community-based tradition stresses the wider involvement and empowerment of various actors, especially host communities, in development by emphasising the elements of social capital in a local context (see Jones, 2005). All these perspectives have their advantages, but also limitations and different outcomes if utilised in (sustainable) tourism development processes.

The resource-based tradition reflects discussions concerning the limits to growth and the existence of a certain objective limit that cannot be overstepped or negotiated. In tourism discussions, it has been based on a theoretically problematic surface of carrying capacity studies focusing on local site scale processes. Although the resource-based tradition does not necessarily mean that developmental impacts are excluded, it may still be too restrictive a foundation on which to define the limits of growth for a form of sustainable tourism that would also be economically viable. In addition, as stated by Erlet Cater (1993: 89), 'true sustainability includes the human dimension,' which means the use of natural capital in tourism activities and often major changes in the environment. Owing to the limited focus of resource-based carrying capacity, McCool and Lime (2001: 385) have even suggested that 'it is now time to bury the concept of numerical tourism and recreation carrying capacity'.

In contrast, the activity-based tradition demonstrates the present and relatively widely accepted idea of sustainability in tourism. From that perspective sustainable tourism can be defined broadly as 'tourism which is economically viable but does not destroy the resources on which the future of tourism will depend, notably the physical environment and the social fabric of the host community' (Swarbrooke, 1999: 13). Definitions like this emphasise the needs of the industry and sustainable use of its resources (Hardy *et al.*, 2002), and instead of being subjects and actors the environment and local communities become the resources. This approach can still reflect the idea that tourism is a tool for sustainable development, but at the same time it strongly represents the industry's perspective, from which growth and its needs are conditions for justifying sustainability: the objective and driving force is to sustain tourism and its resource base for the future needs of the industry. Therefore, some researchers prefer to use the term sustainable development in tourism (Butler, 1999; see Hall & Lew, 1998; Wall, 1996), which involves the ethical aspects of the ideology of sustainability and does not necessarily refer to a tourism-centric approach in development discussions and practices.

References

Anderson, D. and Brown, P. (1984) The displacement process in recreation. *Journal of Leisure Research* 16 (1), 61–73.

Aronsson, L. (1994) Sustainable tourism systems: The example of sustainable rural tourism in Sweden. *Journal of Sustainable Tourism* 2 (2), 77–92.

Askeeke, J.W. and Katjiuongua, O. (2007) *Development of Sustainable Tourism Country Report: Namibia*. Windhoek: Fenata & Nacobta.

Baum, T. (1998) Taking the exit route: Extending the tourism area life cycle model. *Current Issues in Tourism* 1 (2), 167–175.

Bianchi, R. (2004) Tourism restructuring and the politics of sustainability: A critical view from the European Periphery (The Canary Islands). *Journal of Sustainable Tourism* 12 (4), 495–529.

Bryant, R. and Wilson G.A. (1998) Rethinking environmental management. *Progress in Human Geography* 22 (3), 321–343.

Buckley, R. (1999) An ecological perspective on carrying capacity. *Annals of Tourism Research* 26 (4), 705–708.

Buckley, R. (2003) Ecological indicators of tourism impacts in parks. *Journal of Ecotourism* 2 (1), 54–66.

Burns, P. (1999) Paradoxes in planning: Tourism elitism or Brutalism? *Annals of Tourism Research* 26 (2), 329–348.

Butler, R. (1980) The concepts of a tourist area cycle of evolution: Implications for management of resources. *Canadian Geographer* 24 (1), 5–12.

Butler, R. (1991) Tourism, environment, and sustainable development. *Environmental Conservation* 18 (3), 201–209.

Butler, R. (1996) The concept of carrying capacity for tourism destinations: Dead or merely buried? *Progress in Tourism and Hospitality Research* 2 (3–4), 283–293.

Butler, R. (1997) Modelling tourism development: Evolution, growth and decline. In S. Wahab and J. Pigram (eds) *Tourism, Development and Growth: The Challenge of Sustainability* (pp. 109–125). London: Routledge.

Butler, R. (1999) Sustainable tourism: A state-of-the-art review. *Tourism Geographies* 1 (1), 7–25.

Butler, R. (2004) The tourism area life cycle in the twenty-first century. In A. Lew, M. Hall and A. Williams (eds) *A Companion to Tourism* (pp. 159–169). Oxford: Blackwell.

Cater, E. (1993) Ecotourism in the Third World: Problems for sustainable development. *Tourism Management* 14 (1), 85–90.

Clarke, J. (1997) A framework of approaches to sustainable tourism. *Journal of Sustainable Tourism* 5 (2), 224–233.

Collins, A. (1998) Tourism development and natural capital. *Annals of Tourism Research* 26 (1), 98–109.

Craik, J. (1995) Are there cultural limits to tourism? *Journal of Sustainable Tourism* 3 (1), 87–98.

Day, G. (1981) The product life cycle: Analyses and applications issues. *Journal of Marketing* 45 (1), 60–67.

Department of Environmental Affairs and Tourism (2002) National responsible tourism development guidelines for South Africa (Provisional guidelines). Pretoria: Department of Environmental Affairs and Tourism.

Duffy, R. (2002) *A Trip too Far: Ecotourism, Politics and Exploitation*. London: Earthscan.

Ferreira, S.F. and Harmse, A.C. (1999) Social carrying capacity of the Kruger National Park, South Africa: Policy and practice. *Tourism Geographies* 1 (3), 325–342.

Fraser, A. (2007) Land reform in South Africa and the colonial past. *Social and Cultural Geography* 8 (6), 835–851.

Getz, D. (1983) Capacity to absorb tourism: Concepts and implications for strategic planning. *Annals of Tourism Research* 10 (2), 239–263.

Gössling, S. and Hall, C.M. (2005) An introduction to tourism and global environmental change. In S. Gössling and C.M. Hall (eds) *Tourism and Global Environmental Change* (pp. 1–33). London: Routledge.

Government of Botswana (2002) Botswana National Ecotourism Strategy (BNES). Government of Botswana, Gaborone.

Hall, C.M. (1994) *Tourism and Politics: Policy, Power and Place.* Chichester: Wiley.

Hall, C.M. and Lew, A.A. (1998) The geography of sustainable tourism: Lessons and prospects. In C.M. Hall and A.A. Lew (eds) *Sustainable Tourism: A Geographical Perspective* (pp. 199–203). New York: Longman.

Hammitt, W. and Cole, D. (1987) *Wildland Recreation: Ecology and Management.* New York: Wiley.

Hardy A., Beeton, R. and Pearson, L. (2002) Sustainable tourism: An overview of the concept and its position in relation to conceptualisations of tourism. *Journal of Sustainable Tourism* 10 (3), 475–496.

Holden, A. (2003) In need of new environmental ethics for tourism. *Annals of Tourism Research* 30 (1), 94–108.

Holden, A. (2007) *Environment and Tourism.* London: Routledge.

Hughes, G. (1995) The cultural construction of sustainable tourism. *Tourism Management* 16 (1), 49–59.

Hughes, G. and Furley, P. (1996) Threshold, carrying capacity and the sustainability of tourism: A case study of Belize. *Caribbean Geography* 7 (1), 36–51.

Jones, S. (2005) Community-based tourism: The significance of social capital. *Annals of Tourism Research* 32 (2), 303–324.

Kieti, D. and Akama, J. (2005) Wildlife Safari tourism and sustainable local community development in Kenya: A case study of Samburu National Reserve. *Journal of Hospitality and Tourism* 3 (2), 71–81.

Krippendorf, J. (1982) Towards new tourism policies. *Tourism Management* 3 (1), 135–148.

Krüger National Park Management Plan (2006) On WWW at http://www.sanparks.org/conservation/park_man/kruger.pdf. Accessed 15.4.2008.

Lindberg, K., McCool, S. and Stankey, G. (1997) Rethinking carrying capacity. *Annals of Tourism Research* 24 (3), 461–465.

Liu, Z. (2003) Sustainable tourism development: A critique. *Journal of Sustainable Tourism* 11 (4), 459–475.

Lucas, B. (1964) Wilderness perception and use: The example of the boundary waters Canoe area. *Natural Resource Journal* 3 (4), 394–411.

Macbeth, J. (2005) Towards an ethics platform for tourism. *Annals of Tourism Research* 32 (4), 962–984.

Martin, B.S. and Uysal, M. (1990) An examination of relationship between carrying capacity and the tourism life cycle: Management and policy implications. *Journal of Environmental Management* 31 (2), 327–333.

Massyn, P. J. (2007) Communal land reform and tourism investment in Namibia's communal areas: A question of unfinished business. *Development Southern Africa* 24 (3), 381–392.

Mathieson, A. and Wall, G. (1982) *Tourism: Economic, Physical and Social Impacts.* New York: Longman.

McCool, S. and Lime, D.W. (2001) Tourism carrying capacity: Tempting fantast or useful reality. *Journal of Sustainable Tourism* 9 (3), 372–388.

McKercher, B. (1993) Unrecognised threat to tourism: Can tourism survive sustainability. *Tourism Management* 14 (1), 131–136.

Ministry of Environment and Tourism (2005a) A National Tourism Policy for Namibia. Windhoek.

Ministry of Environment and Tourism (2005b) On WWW at http://www.met.gov.na/programmes/cbnrm/cbtourism_guide.htm. Accessed 21.1.2008.

Mowforth, M. and Munt, I. (1998) *Tourism and Sustainability: A New Tourism in the Third World.* London: Routledge.

Murphy, P. (1983) Tourism as a community industry. *Tourism Management* 4 (3), 180–193.

Murphy, P. (1988) Community driven tourism planning. *Tourism Management* 9 (2), 96–104.

Pigram, J. and Jenkins, J. (1999) *Outdoor Recreation Management*. London: Routledge.

Pilanesberg Management Plan (2nd edn) (2006) On WWW at http://www.pilanesberg-game-reserve.co.za/management/index.html. Accessed 17.4.2008.

Redcliffe, M. and Woodgate, G. (1997) Sustainability and social construction. In M. Redcliffe and G. Woodgate (eds) *The International Handbook of Environmental Sociology* (pp. 55–67). Cheltenham: Edward Elgar.

Robinson, G. and Hall, D. (2000) The community: A sustainable concept in tourism development. In D. Hall and G. Robinson (eds) *Tourism and Sustainable Community Development* (pp. 1–13). London: Routledge.

Saarinen, J. (2006) Traditions of sustainability in tourism studies. *Annals of Tourism Research* 33 (4), 1121–1140.

Saarinen, J. and Niskala, M. (2009) Selling places, constructing local cultures in tourism – the role of Ovahimbas in Namibian tourism promotion. In P. Hottola (ed.) *Tourism Strategies and Local Responses in Southern Africa*. Wallingford: CABI Publishing (forthcoming).

Scheyvens, R. (1999) Ecotourism and the empowerment of local communities. *Tourism Management* 20 (2), 245–249.

Scheyvens, R. (2002) *Tourism for Development: Empowering Communities*. Harlow: Prentice-Hall.

Sharpley, R. (2000) Tourism and sustainable development: Exploring the theoretical divide. *Journal of Sustainable Tourism* 8 (1), 1–19.

Stankey, G. (1982) Recreational carrying capacity research review. *Ontario Geography* 19 (1), 57–72.

Stansfield, C. (1978) Atlantic city and the resort cycle: Background to the legalization of gambling. *Annals of Tourism Research* 5 (2), 238–251.

Stronza, A. (2007) The economic promise of ecotourism for conservation. *Journal of Ecotourism* 6 (3), 210–230.

Swarbrooke, J. (1999) *Sustainable Tourism Management*. Oxon: CAB International.

Taylor, J.P. (2001) Authenticity and sincerity in tourism. *Annals of Tourism Research* 28 (1), 7–26.

Telfer, D. and Sharpley, R. (2007) *Tourism and Development in the Developing World*. London and New York: Routledge.

Timothy, D. and White, K. (1999) Community-based ecotourism development on the periphery of Belize. *Current Issues in Tourism* 2 (2), 226–242.

Tooman, L. (1997) Applications of the life-cycle model in tourism. *Annals of Tourism Research* 24 (1), 214–234.

Vaske, J., Shelby, B., Graefe, A. and Heberlein, T. (1986) Backcountry encounter norms: Theory, method and empirical evidence. *Journal of Leisure Research* 18 (2), 137–153.

Wall, G. (1982) Cycles and capacity: Incipient theory of conceptual contradiction? *Tourism Management* 3 (3), 188–192.

Wall, G. (1996) Is ecotourism sustainable? *Environmental Management* 2 (2), 207–216.

Wagar, A. (1964) The carrying capacity of wild lands for recreation. *Forest Science Monographs* 7 (1), 1–24.

Wheeller, B. (1993) Sustaining the ego. *Journal of Sustainable Tourism* 1 (1), 121–129.

WTO (1993) *Sustainable Tourism Development: Guide for Local Planners*. Madrid: Word Tourism Organization.

Part 2

Tourism Development and Local Policies of Sustainability

Chapter 6

Tourism, Conservation Areas and Local Development in Namibia: Spatial Perspectives of Private and Public Sector Reform

FRITZ BECKER

Introduction

Since its birth in 1990, the Government of the Republic of Namibia pursued a broad reform of the private and public sector in the country, which included the tourist business and industry. A rather comprehensive restructuring process of the *Ministry of Environment and Tourism* (MET) widened the ministerial core business, focusing on 'environment and tourism' as an agent for growth and development. The deregulation of exclusive Apartheid avenues to means of material and non-material (re-) production in tourism challenged all actors propelling the reformatory efforts. They created opportunities for national economic growth, individual education, training and employment. The spatial projection of developments on communal lands, arid landscapes and remote communities generated a participatory system of expanded regional, local and individual benefits, rights and responsibilities. Simultaneously, the countrywide implementation of concepts, programmes and projects pertaining to the development of tourism and protective conservation of natural heritage, triggered the transformation of land-use patterns and of classified protection areas.

This retrospective chapter reflects on effects and spatial projections observed in the reform of 'tourism and environment' that ties and strengthens the triangular relationship between

(1) progresses in sustainable tourism;
(2) conservation and protection of fragile ecosystems; and
(3) the mindful integration of local communities into regional planning.

The author offers aspects on the role that one of Namibia's export sectors plays in the spatial transformation of the country, aiming at distributive access to wealth and human quality of life through tourism development with environmental governance.

Tourism Reform in Flux

In 2008, one may conclude that Namibia's Government completed the post-colonial mission to prepare the country's natural environment and tourism potential for future exploitation in a sustainable manner. In noting the fact that the MET is awaiting the protracted passing of a bill into law, and that sister ministries are busy reviewing impacts of their earlier reformatory decisions may indicate that retardation and change constitute elements implicit in the tourism reform. Over the years, elements such as 'gained experience' and 'built capacity' surfacing from the reform process, convinced private and public institutions to re-assess the effectiveness and efficiency of their revised institutional organisation and structures, realising the necessity to rectify emotional and material consequences following implemented changes.

Problem formations

In times of change, it is almost inevitable to observe that transformation and transition produce emotional interfaces among interested groups. In the Namibian arena of sustainable tourism development with (or versus?) nature conservation, the local level of community-based tourism (CBT) continues to form interfaces. The conflict formation – counter-productive to swift development – usually evolves from (in-) formal practices and asymmetries applied to the distribution of material benefits or the sharing of access to material (re-) production. In the context of rural lands, Seely and Zeidler (2002: 79) exemplify conflict over land distribution and land-use pointing out the individual disregard for and even violation of the spirit carried in collective principles such as 'equity' and 'equality of opportunity'. Long (2004: 131) confirms this problem in his assessment of livelihoods and tourism in communal areas.

Research and monitoring

In cooperation with their parastatals, all ministries and most tourism enterprises are both the targets of reform as well as the driving agents of change. Conscious of this dual role, the MET closely monitors the quality of the transition. It evaluates the impact of the transformation pertaining to the institution, the growth of the tourism economy, developments in the tourism markets and emerging patterns of individual or collective land-use and ownership.

Focusing on Namibia, national institutions of tertiary education and research augment(ed) their studies into this functional relationship. They analyse the structural and horizontal transformation that characterises this pivotal component of the country's export economy. Progress reports and reviews published by the 'Namibian Association of Community-based Natural Resource Management (CBNRM) Support Organisation' (NACSO) give a remarkable quantitative account of instrumented progress in pro-poor livelihood rehabilitation, directed towards people living in communal areas. Vorlaufer (2007: 29) indicates that – along with NACSO – non-governmental organisations (NGOs) such as the Namibia Community-based Tourism Association (NACOBTA) and the project conducted by the World Wildlife Fund – Living in a Finite Environment (WWF-LIFE) offer(ed) their assistance to the careful creation of avenues to income generation in support of the 'laboratory of reforms' in the tourism sector.

The spatial projection

Commonly, people perceive directions and effects of ministerial reforms being rather vertical than horizontal in nature. The MET projects its sector activities such as 'conservation' and 'tourism growth and development' on 'earth-space'. The proclamation of 50 conservancies over a period of nine years and 13 community forests in 2006 (NACSO, 2007: 16) testifies the Ministry's determination to transform 'space' into environments with prospects for poor people struggling in regional survival economies. The establishment of conservancies and community forests couples with the introduction of prominent development concepts coined as 'CBT' and 'community-based natural resource management'. Their registration as spatial entities proves the consumption of 'surface area', furnished with classified permissions governing the utilisation of natural resources. It entails the restructuring of territorial as well as social spaces and includes the endeavour to (re-) introduce perhaps lost, indigenous land-use practices and income opportunities, locally.

At the international level, the country's commitment to the spatial system of trans-frontier conversation demonstrates another quality of domain the State and MET covers (see Scholz, Chapter 9 in this book). At sub-national level, the awaited promulgation of the 'Parks and Wildlife Management Act' (PWMA) will likely define six different levels of conservation areas on communal and commercial (farm) land (Table 6.1). The interpretation of the Act will add an important planning tool to the governance of spatial projections that emerge from regional tourism planning in the north-west, north-east and south of Namibia, where according to Long (2004) examples of integrated, regional tourism development plans have been launched.

Table 6.1   Categories of protected areas in Namibia (proposed)

Category	*Definition*
Wilderness area	To maintain its outstanding or representative ecosystems, geological or physiological features or species in an undisturbed state primarily for the purposes of scientific research or environmental monitoring, and to provide limited access to areas that have been largely undisturbed by human actions over a long period of time
National park	To protect the ecological integrity of one or more ecosystems for present and future generations, to exclude exploitation or occupation inconsistent with such protection and to provide a foundation for compatible spiritual, scientific, educational, recreational and visitor opportunities
Natural monument	To protect or preserve in perpetuity a specific outstanding natural feature or features, which is (or are) of outstanding or unique value because of its inherent rarity, representative or aesthetic qualities or because of its cultural significance
Nature reserve	To secure and maintain by active intervention for management purposes, areas that are necessary to ensure the maintenance of habitats and/or to meet the requirements of specific species and to facilitate scientific research, environmental monitoring and public education relating to sustainable resource management
Protected landscape	To maintain the harmonious interaction of nature and culture, through the protection of areas where the interaction of people and nature over time has produced an area of distinct character with significant aesthetic, ecological and/or cultural values, and to promote the continuation of traditional human practices in such areas
Conservation area	To ensure long-term protection and maintenance of its biological diversity while providing at the same time a sustainable flow of natural products and services to meet community needs

Source: MET. Parks and Wildlife Management Act. *Working Document.* May 2001, 23–24. (Unpublished manuscript; see also: Barnard, 1998: 295–296). Tabularisation: F. Becker, 12/2007

Creating Space

These instituted commitments indeed translate into the many processes pertaining to local and regional planning of space. They regulate the national territory and the 'production of social spaces' (Becker, 2006a: 19–20). It is important to acknowledge that the second and third National Development Plan pay increasingly attention to matters of comprehensive and structured spatial (development) planning (Republic of Namibia. National Planning Commission, 2001, 2007). The MET's key areas of business act complementary to incremental planning approaches applied by governmental administrations that respond to rising and extending land claims in conjunction with emerging rural–urban settlement systems of different types and prospects in communal areas. On the other hand, they serve the export demand for non-renewable raw materials, when demarcating mining concession areas for foreign investors in the light of booming strategic minerals' markets.

Paradigms, instruments and ethics

The negotiation and delineation of conservancies for CBT in public–private partnership with the tourism industry constitute favoured instruments employed in rural tourism development, and enhance the Ministry's in-house expertise in local management issues. The expertise on site assists in shaping the territorial planning of land-use and land consumption. Prominent paradigms, policies, instruments and objectives underlie the implementation of (spatial) development planning, reflecting ethic principles captured in terms such as

(1)   sustainability;
(2)   biodiversity;
(3)   participation; and
(4)   equity and environmental justice.

Responsibilities and obligations inherent in these principles demand firm courage from professionals who want to live through their convictions. It is no secret that decision-makers in spatial planning and natural resource development, who subscribe to such (perhaps imperative) human behaviour, face almost daily the contradiction from the camp of the 'economic man' in sector and spatial planning. From this perspective, decision-making strives to achieve monetary targets of economic growth with low levels of concern for unsustainable environmental impact; seemingly, always willing to ignore planning instruments like the environment impact assessment (EIA).

The reform ultimately determines the integration of marginalised land and people into the national economy. Its spatial projection substantiates the complexity and strain that accompany 'cause and effect' relationships, resulting from the creation and development of spatial categories such as conservancies or protected areas. The MET strongly pursues the facilitation of sustainable (eco-) tourism development on remote lands (see Scholz, Chapter 9 in this book). The economic intent of such planning turns (semi-) arid landscapes into commercial commodities that transform people's social practice, affecting the socio-cultural identity of all involved.

The underlying processes of this actor-based creation of space construct new and competing 'spatialities' and do not reflect 'territoriality'. Becker (2006a: 19–20) argues that spatial action delimitates 'social space', and the process often harbours politico-economic and socio-cultural problem formations. The implementation of CBT in communal conservancies usually reveals, if not nurtures, conflict resulting from uneven access to means of production like land. Such tensions in customary societies easily evolve from contradictory individuals or collective perceptions on 'culture and land' as Chirawu (2004: 237) describes. However emotional, Hinz (2002: 197–207) exemplifies the deliberate planning behind conflict formations that sometimes tell the story how rational informed people choose either traditional or general law in order to secure profits and, finally, defend their action space (spatiality) in tourism landscapes with scarce resources and rare propellant prospects, at first sight.

Pro-poor concern

Deliberations on reforms focusing on tourism-based development would remain incomplete without a brief reference to academic concern over hailed concepts and instruments mirrored in many tourism projects. Some discourse centres around dissatisfaction evolving from *ex post* project monitoring that neglects the discussion of missing pro-poor impact felt in communities hosting tourists. Questions are essentially seeking answers to the (iterative) observation, why the 'tool box' of sector and spatial development (e.g. concepts; instruments) perpetuates disregard for pro-poor design and methodology; not being culture conscious and enabling or, participatory and distributive. In the pursuance of people-centred transformation from fragile local conditions of basic human needs towards improved human qualities of life, the challenge continues to set the lyric apart and experiment with – perhaps innovative – basic and application-oriented methods and techniques in pro-poor field research and planning of tourism areas (see Becker, 2006b).

Tourism Growth and the Discovery of Landscapes

At the turn of the millennium, the World Travel & Tourism Council (WTTC) predicted a breath-taking increase in foreign tourists entering Namibia, covering the decade 2000–2010 (WTTC, 2000). The forecast presents promising estimates on primary and secondary employment effects including economic growth that invite a thorough analysis of possible regional disparities in future, taking into consideration that many imponderable factors influenced the validity of statements during the course of the past eight years (see Becker, 2002: 69).

Expectations and aspirations

It seems to be expedient to offer a few challenging expectations and aspirations. Becker (2003: 207) outlined challenges underlying their possible materialisation, which require foresighted action from the Namibian tourism sector in view of the magnitude their management entails. In 2007, experts in the tourism business, for example,

(1) anticipated that Namibia will receive one million visitors in 2008;
(2) emphasised that domestic tour products and accommodation establishments meet any taste and budget;
(3) discussed that more up-market hotels will be built in Windhoek and on the coast;
(4) hoped that the World Soccer Championships 2010, convened in South Africa, will attract event tourists from overseas to Namibia;
(5) estimated that forty of hundred jobs will be (in-) directly linked to tourism by 2010; and
(6) assumed that the tourism sector's share in Namibia's gross national product will exceed 13%.

Shaw and Williams (1994: 95–153) exemplified backward and forward linkages, the consumptive nature of tourism triggers in all sectors of the economy. The anticipated boom and the competition felt among SADC members virtually mandate the national ministerial offices and private companies concerned to enhance capacities and undertakings in their fields of competence (see Becker, 2003: 207). Astronomical investments in tourism accompany this successful endeavour. The investors expect that they facilitate the reduction of imports and further domestic supply markets, producing Namibian goods. Such challenges send terrifying warning signals in anticipation of pressures likely imposed on the society and resources, although the export industry's future appears to look quite promising. One of the warning signals doubts the public spiritedness of the industry's beneficiaries. They are tasked iteratively to ensure that the (re-) distribution and (re-) investment of profits and earnings will also serve the 'alleviation of poverty of the poor', irrespective where the struggle for survival of disadvantaged is hardest in the country. Another signal highlights people's hope that Namibia's negative resource transfer will be minimal for the sake of sustainable development (see Becker, 2002: 70). Both aspects are acknowledged prerequisites for stability and peace at any tourism destination.

Branding, tourism landscapes and images

Scholz (Chapter 9 in this book) presents the kaleidoscope of legislative, institutional, conceptual and instrumental measures the MET undertook and pursues in order to put the developmental direction and impact of 'environment and tourism' on a solid foundation. In the domain of tourism marketing, the definition of the assemblage of the country's landscape(s) supported the branding of Namibia as a tourist destination.

The making of the sense of place and landscapes led to the sale of the imagined last refuge of wide, open, wild and uninhabitable spaces.

The creation of landscapes for the purpose of tourism reminds geographers of the pivotal role this 'polysemic term referring to the appearance of an area, the assemblage of objects used to produce that appearance' (Duncan, 1994: 16) played in the discipline's conceptual interpretation, scientific perception and research fields in the twentieth century. After more than three decades of shadowy existence, this subject's core returned to application-oriented geography, for instance, in the work of 'landscape ecology'. The discovery of the geographic concept of landscape (=Landschaft in the German geography tradition) for marketing purposes in tourism rather intends to capture the perception of landscape (Landschaft) that the Finnish geographer Granö published in 1929. In his definition, 'landscape includes both the combination of visual phenomena and the complex of objects perceivable through these visual phenomena (...)' (Granö, 1997: 50). This understanding underlies the creation of the Namibian tourism landscapes.

Connected to the various objectives and targets of tourism development and environmental protection, landscapes inspire numerous professions to apply the concept's subjective and interpretative virtues. The MET's protection areas strive to sustain the geo-ecological, economic, cultural and aesthetic values of Namibia's flora and fauna. Whenever necessary, the Ministry rehabilitates endangered biota and their landscapes. Despite 'nature' of the biome, biota, including people, constitutes commodities, which the creative mind of the 'economic man' turns into marketable products of profitable value. The printing press and the media at large specialised in tourism, offer visitors glamorous books in colour print and films that invent natural and cultural phenomena of the country's landscapes. The iconography of such publications seeks to attract tourists with the myth of the depicted features. The 'visual narrative' draws individual 'mental maps' of 'nature, people and culture'. It nurtures tourists' desires to flee their everyday life in order to experience the virtual perception of otherness in 'reality'. With reference to McIvor (2000: 19f), it means to engage in trophy hunting, photography, learning or in ecotourism and receive stimulating recreation from enjoying the horizons, heritage (Box 6.1) and cultures of Namibian landscapes. A number of geographic publications characterise a selection of Namibia's arid regions, the salient features of which Hüser *et al.* (2001) offer the traveller en route. Geographers interpret many of these works as landscape science (Landschaftskunde), simultaneously explaining this geographic concept to the public, in general. These detailed and knowledgeable publications mainly highlight the Namib Desert, the Central Plateau, the Savannahs or the Kalahari. The extraordinary landscapes invite tourists from overseas to prepare for the envisaged safari or to read the excursion again, at home.

Box 6.1 Cultural Heritage Tourism and Sustainability in South Africa: The Case of the Cradle of Humankind

FELICITE A FAIRER-WESSELS

Cultural heritage tourism can be seen as 'an experience which is produced by the interaction of the visitor with the resource' (Moscardo in Apostolakis, 2003: 799) and includes the travelling to experience the places and activities that authentically represent the stories and people of the past and present (National Trust for Historic Preservation, 2008). Cultural heritage tourism can encourage and result in sustainable economies in especially rural communities if the cultural, historic and natural resources are managed, protected and preserved authentically; and the needs of the communities be balanced with those of tourists.

Although heritage and history are often used interchangeably, they are two different constructs. Heritage can be described as history which has been processed through mythology, ideology, nationalism, local pride, romantic ideas, or marketing, into a commodity. Heritage is about how we mediate our relationship with the past, whereas history is concerned with the causes and effects of changes over time. Peoples' expectations of heritage sites are somewhat influenced by the way in which the mass media portrays heritage or history (Herbert, 1995: 21, 26, 147; Prentice *et al.*, 1998: 1–25).

Heritage is our legacy. It is what we live with today, and what we pass on to our future generations. Our cultural as well as natural heritages are both irreplaceable sources of inspiration (see South Africa's World Heritage Sites, 2007). People's motivation to travel appears to be changing. More people visit heritage sites, namely those with a genuine interest in history and a good education understand the significance of cultural heritage attractions (Poria *et al.*, 2006: 319).

In recent years, there has been a convergence between travel and natural and cultural heritage. A vast amount of change has taken place in both production and consumption patterns underpinning heritage tourism. The typical tourist is moving away from the sun- and sea-type holiday or vacation, towards a more sophisticated type of holiday containing unique personal experiences, as well as exclusivity and differentiation. Producers and consumers are thus turning their attention to heritage tourism. Heritage attractions have therefore developed a niche in the industry and are now regarded as a contemporary commodity created for the purpose of satisfying contemporary consumption (Apostolakis, 2003: 796). As more people seek an experience rather than just a day's outing, heritage attractions now have a new niche market to target.

Over the years, many heritage attractions have been established all over the world. Heritage sites are currently competing for a portion of a relatively stable market due to the fact that the supply of heritage attractions now exceeds the current demand. Although the number of visitors to heritage sites has increased, demand for each resource has actually decreased. That is why it is important to offer a product that satisfies visitor expectations (Apostolakis, 2003: 798).

World heritage sites can boost local economies by encouraging tourism. For these reasons, world heritage sites need to be preserved for future generations, as part of a common universal heritage. A large amount of prestige is brought to a site which has world heritage status as it may help to promote the site, both locally and internationally, as well as attract new visitors. The highest quality standards are encouraged regarding the management of such sites, which means international accountability, and if a site is threatened it can be added by the World Heritage Committee to the List of World Heritage Sites in Danger.

The Cradle of Humankind was nominated by South Africa as a world heritage site in 1998 and declared a world heritage site of cultural significance by United Nations Educational, Scientific and Cultural Organisation (UNESCO) on 2 December 1999. In 2005, UNESCO extended the appellation 'Cradle of Humankind' to include two additional locations that are conceptually connected to the Sterkfontein area but not geographically contiguous to the original site. These sites are the Taung Quarry near Kimberley and Makapansgat near Mokopane. Both these sites consist of broadly the same kind of rock desposits as Sterkfontein and both contain valuable fossil specimens, which confirm that hominids have lived in southern Africa for an unbroken stretch of nearly 3.5 million years. The Cradle of Humankind, situated mainly in the North West Province, covers an area of 47,000 ha. This site has been described as the place where humankind began. Due to its unique nature it has established itself as a site of universal value (Fleminger, 2006).

The Cradle of Humankind comprises a band of around 40 important palaeo-anthropological sites, 13 of which have been excavated including Sterkfontein, Swartkrans, Kromdraai, Coopers B, Wonder Cave, Gondolin and various others.

The significance of this area lies in the fact that the Cradle of Humankind is a scientific treasure chest of information on human and cultural evolution. *Australopithecus*, which is one of the fossil species found at this site, has been recognised as a distant relative of all humankind. It is claimed to be a small-brained hominid, an early human ancestor, which walked on two legs. Two well-known examples of this hominid are the skull of Mrs Ples and a nearly complete skeleton called Littlefoot, which were discovered in the area in 1947, with the Taung skull already identified in 1924 (Fleminger, 2006). The Cradle of Humankind is by far one of the world's most important sites for researchers and human evolutionary studies (South African Places, 2006).

Tourism as a powerful economic development tool has presented many business opportunities with the Cradle of Humankind Management Authority doing everything in its power to involve, uplift and boost local interest through community projects and job creation. In addition, community development and educational trusts have been established. This has improved the quality of life for the resident communities and has helped in protecting South Africa's natural and cultural treasures. An important benefit of cultural tourism is that opportunities increase diversified economies, ways to prosper economically while holding on to the unique characteristics that distinguish communities.

With the recent surge in tourism, both local and international, the various authorities responsible for The Cradle of Humankind took a bold step in creating a new and unique visitors centre that acts as a draw card for the area – 'Maropeng', meaning 'returning to the place of origin'. Maropeng cost around R163 million to build and caters for up to 600,000 visitors per year, while local infrastructure was also improved at the cost of R180 million (Fleminger, 2006: 67). Maropeng Interpretation Centre features novel exhibitions for the preservation and the interpretation of the site to teach visitors their importance and, by extension, the importance of preserving other such sites elsewhere.

The Cradle of Humankind and Maropeng use traditional media and internet marketing to market their cultural heritage to attract new audiences. The Cradle of Humankind is a site that has developed its potential for cultural heritage to create new opportunities for tourists to gain an understanding of an unfamiliar place, people and time.

The positive collaboration between tourism and heritage and culture can go a long way in promoting local economies, with the focus of cultural heritage tourism working towards sharing heritage and culture with visitors in order to reap economic benefits. This also promotes community pride as people work towards a common goal in establishing a thriving tourism industry in South Africa.

Sustaining commercialised landscapes

Namibia's private and public tourism industry commercialises the branded, eye-catching sceneries of the country, globally. The marketing of the fragile ecosystem, the natural commodity or the tourist destination 'Namibia' requires quite an entrepreneurial effort. The continuation of the product's life cycle depends on the informed analysis of old and emerging international tourist markets and holiday-makers' changing preferences for destinations with reasonably priced tour packages. Negative experiences recorded worldwide, report the depletion of resources, pauperisation of local communities and loss of cultural identity following tourism growth and development of wasteful exploitation (see also Mabunda & Wilson, Chapter 7 in this book).

In Namibia, the MET strives to bind the instrument of EIA to the process of single sector and spatial planning in order to minimise unwanted developments. Crucial factors investigated in the planning process refer to (i) the fragility of geo-ecosystems; (ii) availability of potable water and (iii) carrying capacity of the environment concerned (Becker, 2002: 70). The instruments of the EIA are set to prevent any uncontrolled consumption of land and landscapes. The public seems to observe occasionally that – perhaps with tacit approval – investors disregard and violate the legislation in place for the protection of the environment, embedded in a carefully structured and comprehensive planning process (Shaw & Williams, 1994: 234). Subsequently, the Ministry attempts in an envisaged move to eliminate marked natural and cultural landscapes from investments that do not comply with the standards of the EIA; properties that would qualify affected landscapes and regions for protection appear to remain undefined, currently.

Landscapes, Spatial Structures and Land-use in Transformation

From Independence, the MET's work proves an admirable persistence in pursuing environmental concerns over resource and landscape conservation and protection. The ministerial performance represents political pressure and governmental advocacy, which 'one office' disseminates simultaneously. This dual practice is enshrined in the political ecology of competition with exponents of mechanistic economic growth. Within this politico-economic framework and professional call for the sustainable benefit of the country's people, the Ministry's departments continue to urge on the consolidation of existing, in part colonial laws by propelling the elaboration of an adequate legislation pertaining to the Ministry's portfolio. The results implicate to serve the interest of the State and regulate society in the sharing consumption of all means of (re-) production in Namibia.

An outstanding document of legislation characterising the described situation constitutes the pending PWMA, awaiting promulgation. Until passed into law, the MET governs principles of the forthcoming PWMA with the assistance of the Nature Conservation Amendment Act (1996). Becker (2002: 72) underlines the strategic significance of this amended Act. It provides an opportunity for development triggered 'from below' in conservancy areas on communal lands; including the right to hunt game for the benefit of the local communities. This progress equalises the status that tenants of commercial farmland used to enjoy exclusively since 1967.

Spatial planning through parks and wildlife management

The PWMA facilitates the dividing of vast areas of the national economic territory into six spatial units with distinct borders, differing functional appropriation and intensity of utilisation. The definitions of all the six protected entities subscribe to the philosophy of sustainability. The management of these categories is bound to individual objectives relating to conservation and protection of their geo-ecological conditions and potentials. It may be important to highlight that the definition of each spatial category conditions (i) the utilisation for research and education, (ii) the economic exploitation or (iii) the cultivation of heritage and indigenous values (see Table 6.1). Applications seeking individual permission to enter the areas under protection will probably examine the interpretation of the law by practice under the then prevailing social and political conditions. Any party holding a permit to enter and use the defined spatial categories will have to accept the EIA procedure and control. The public will likely observe, by nature, tangible problems evolving from the contestable interpretation of the framework law. Any kind of implementation of this Act indeed might reveal flexible characteristics of this Act; a condition usually noted in spatial planning legislation.

The definitions reasoning the units' classification of protection intentionally exclude any indication or prescription concerning their spatial extent and avoid any attempt to typify the inner structure of the surface areas; for example, by naming binding components of typology or ecology (see Table 6.1). Upon granting a permit, the protected area proclaimed turns virtually into a delimited unit for sector and spatial planning of local and regional scale. Any private or public permit holder is obliged to look after the unit in accordance with the permission stipulating its utilisation; property rights on land remain untouched.

Spatial structures in transition

Any proclamation of protected areas provides the opportunity to apply the interpretation of this Act in terms of a national planning instrument.

In anticipation of an instituted spatial planning legislation that would direct the consumption and regulate the land-use of Namibian territory in an integrative manner from national to local level, the PWMA could play a pivotal role in the needed national to local planning system. Elements of such a system would comprise (i) national guidelines; (ii) planning legislation and special acts; (iii) advisory comprehensive plans and (iv) legally binding implementation plans. The implementation of the PWMA could form a compound spatial planning system that would mark and connect spatial structures such as

(1) monopolised mining areas like the diamond concession area *Sperrgebiet* south of Lüderitz, proclaimed as National Park in 2008;
(2) tourist attraction areas like the famous Etosha National Park;
(3) emerging settlement patterns; and
(4) cordons of classified buffer zones, separating conflicting land-use patterns.

The rather holistic approach to national spatial planning would put the MET in the position to project a variety of ministerial sector activities on space, similar to the quality of country planning exercised in Botswana. In collaboration with sister administrations and their parastatals, less incremental planning with more cooperation in comprehensive spatial planning would regulate the structural distribution of the economy, society and the consumptive habits towards land, ultimately 'space'. Equally important, an instituted cooperation in spatial national planning could transform the 'culture of communication, the way the parties concerned (e.g. government and private companies to local communities) deal with each other and talk to each other' (Becker, 2006a: 33) about environmental rights, CBT or local infrastructure needs in conservancies (see Becker, 2002: 72; see Ramutsindela, Chapter 10 in this book). The participatory style of communication would likely accelerate the achievement of the canon of tourism development objectives, goals and targets; the implementation of which results in a projection on space generating tourism landscapes. With a view to conflict formations over, for example, customary grazing rights surfacing in the northern Regions in 2006 and 2007, Becker (2002) stated that remembering the culture of 'participatory communication' might open a new chapter in the history of land utilisation and spatial development in Namibia.

Conservation, tourism and spatial development

The realisation of the MET projects in the course of the tourism reform has a notable spatial impact on the various ecosystems and locations in the country's regions. This observation substantiates the line of arguments in favour of comprehensive sector cum spatial (sub-) national development planning, and the implementation thereof. In the period from 1998 until 2006, Namibia registered a total number of 50 conservancies on communal lands in nine regions, of which the administration established 41 conservancies from 2000. At the same time, the number of regions participating in the programme increased from three to nine. The regions Kunene, Caprivi and Otjozondjupa led the country's 13 regions with 34 conservancies registered until 2006.

The conservancies cover the most frequented tourism landscapes of northern Namibia, offering cultural heritage, deserts, savannah, forests, birds and the 'big five'. The southern regions Hardap and Karas complement the increasingly sutured patchwork of communal conservancies only recently. Likewise, the conservancies in the regions Omusati, Oshikoto and Kavango are comparatively young members of the conservancy network. The registration of the King Nehale Conservancy (2005) in the Oshikoto Region correlates with the opening of the north-eastern Etosha National Park gate, bearing King Nehale's name. The delimitated area of all 50 communal conservancies covers 118,704 km^2, equivalent to 16.33% of the nine regions' territorial space (727,129 km^2). The regions Kunene, Caprivi and Otjozondjupa managed to absorb 68% (1998–2006) of the 50 conservancies collated in Table 6.2. In addition to the statistic information on the spatial distribution, the data provide an idea about the tourists' sightseeing preferences in Namibia and, in return, the making and marketing of places at tourist fairs, overseas. The quantitative reference made to the number of people living in a given communal conservancy does not necessarily equate with the number of people registered as members of the particular conservancy.

Despite the fact that conservancies on freehold land dominate the more central part of the country and occupy 6% of Namibia's surface area, the writing will only touch this particular spatial development (see Becker, 2002: 73). Located on commercial farmland, these conservancies stretch from south of Windhoek to the southern fence of the Etosha National Park in the north, and from Khorixas in the west to Tsumeb in the east (Figure 6.1). This type of conservancy supports 'conservation', twinning 'tourism' and 'biodiversity'. It connects the various local to regional land use areas in the country. In terms of spatial development planning, these conservancies fill the puzzle of state-controlled spatial land-use structures and development plans evident through the Protected Area Network (PAN) (see Barnard, 1998: 43–44). The said conservancies' role in the foreign exchange economy of tourism is substantial in comparison with CBT in communal areas, and their contribution to the concept of biodiversity is remarkable. From Independence, the complexity characterising the intimate relationship of the two concepts 'conservancy' and 'biodiversity' (see Albl, 2001) enjoys international research interests, anchored all over the country.

Table 6.2 Conservancies in communal areas in Namibia (2006)

Region	Territory (km²)	Conservancies per region	Period registered	Area (km²)	% of region(s)	Number of people
Kunene	144,255	18	1998–2006	39,697	27.52	30,670
Caprivi	19,532	09	1998–2006	2563	13.12	26,900
Otjozondjupa	105,328	07	1998–2006	43,109	40.93	34,300
Kavango	48,456	04	2004–2005	1190	2.45	6000
Erongo	63,586	04	2001–2006	17,419	27.39	6800
Karas	161,086	03	2003–2005	4627	2.87	10,500
Omusati	26,558	03	2003–2005	9496	35.76	85,360
Hardap	109,659	01	02/2001	95	0.09	120
Oshikoto	38,669	01	09/2005	508	1.31	20,000
09 Regions	727,129	50	1998–2006	118,704	16.33	220,650

Source: NACSO (2007: 12–13). Tabularisation: F. Becker 12/2007

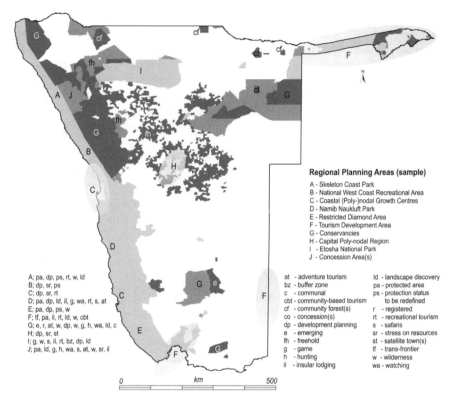

A; pa, dp, ps, rt, w, ld
B; dp, sr, ps
C; dp, sr, rt
D; pa, dp, ld, il, g, wa, rt, s, at
E; pa, dp, ps, w
F; tf, pa, il, rt, ld, w, cbt
G; e, r, at, w, dp, w, g, h, wa, ld, c
H; dp, sr, st
I; g, w, s, il, rt, bz, dp, ld
J; pa, ld, g, h, wa, s, at, w, sr, il

Regional Planning Areas (sample)

A - Skeleton Coast Park
B - National West Coast Recreational Area
C - Coastal (Poly-)nodal Growth Centres
D - Namib Naukluft Park
E - Restricted Diamond Area
F - Tourism Development Area
G - Conservancies
H - Capital Poly-nodal Region
I - Etosha National Park
J - Concession Area(s)

at - adventure tourism
bz - buffer zone
c   - communal
cbt - community-based tourism
cf - community forest(s)
co - concession(s)
dp - development planning
e   - emerging
fh - freehold
g   - game
h   - hunting
il - insular lodging

ld - landscape discovery
pa - protected area
ps - protection status
     to be redefined
r   - registered
rt - recreational tourism
s   - safaris
sr - stress on resources
st - satellite town(s)
tf - trans-frontier
w - wilderness
wa - watching

0   km   500

Figure 6.1 Spatial classification for regional planning in Namibia (example)

Empirical research in conservancies evaluating CBT partnerships entered into in accord with the MET seeks to shed light on the rather asymmetric exchange of goods and services between urban centres housing tourism industries and sparsely populated tourism landscapes. It is assumed that multi-disciplinary field studies will present detailed findings on the question, how the socio-cultural networks of conservancy members (individually and collectively) drive the dual economy of CBT that structures, polarises and sustains the 'rural–urban continuum'.

CBT for Local Development?

The concept of CBT recommends itself almost uncontested as an agent for the development of livelihood, resources and infrastructure in communal areas. This becomes evident when communities enter into cooperation with NGOs and private tourism companies (see Saarinen, Chapters 5 and 15 in this book). The provision of access to the generation of monetary income derived from CBT and other means of income aims at broadening the material and educational resource base of the conservancies. The intent is to guide their members away from the conditions of survival economy and subsistence agriculture towards participation in the regional economic cycles of production, consumption, demand and supply markets that eventually link with domestic urban centres such as Windhoek, Swakopmund or Walvis Bay. The regional integration of islands of CBT and the anchoring of their local nodes pose a major challenge to all institutions and human resources involved in the development of communal conservancies. A challenging task constitutes the provision of basic infrastructure like roads, electricity, potable water, health and schooling in scenic landscapes. Occasionally, it finds itself detached from mainstreams of vertical sector planning and the implementation of comprehensive national and regional planning.

Social mix versus monetary results?

In CBT, the interplay of actors such as public institutions, private partners and members of conservancies tends to build quite heterogeneous communities, lacking bond. Such a scenario in affected communities often results from dominating individual interests and unaccepted ethnic otherness. The condition jeopardises the collaboration of all parties concerned. It determines rather a short than a long period of cooperation, and may confront communal societies, their social coherence and structure of customary law.

The generally favourable opinion on the efficiency of integrated CBT in communal conservancy development raises the question, which parameters trigger growth in income, tourist numbers or game, and which support human education, training or empowerment. In his findings, Long (2004: 119–138) offers quite an exhaustive synopsis that unfolds the connectivity existing between the reciprocal linkages of local livelihoods, tourism, communal areas and conservancies. His conclusions provide a general summary of positive livelihood outcomes from tourism and submit some elements for improvement of livelihood outcomes from CBT (Figure 6.2). Progress data (NACSO, 2007) confirm the monetary value of conservancies' income, amounting to an impressive 19 million Namibia dollars (Table 6.3), and Long (2004: 124) noted that the yield earned from tourism was to be occasionally higher in previous years.

Vorlaufer (2007) harbours doubts about the common assumption that localised CBT directed to the poor in communal areas will alleviate poverty in sustainable terms of monetary income and in human quality of life. In his empirical field study conducted in Namibia, he resumes, for instance, that tourism will not 'generate enough income to satisfy the expectations

			POSITIVE LIVELIHOOD OUTCOMES
Tourism in communal areas is delivering…	Income and employment	▶	
	Enabling environment for local entrepreneurship		More income – individual and collective
	Capacity building; Career development		
	Value to culture		Well being
	Improved governance		Better natural resource management
	Understanding of value of natural resources for tourism		
Elements likely improving livelihood outcomes from CBT…	Create enabling environment for local entrepreneurship	▶	IMPROVED LIVELIHOOD OUTCOMES
	Foster good local governance / leadership		More individual and collective income ↕
	Strengthening legislation – stronger tenure rights to organised communities (e.g. conservancies) to get greater share of tourism value		Better natural resources management

Figure 6.2 Livelihood outcomes from tourism and improvement through CBT
Source: Long (2004: 136–137). Tabularisation: F. Becker, 12/2007

Table 6.3 Conservancy income (2006)

Source of income	_Value (N$)_	_In % of total income_
Joint venture tourism	9,830,198	51.6
Trophy hunting	6,075,868	31.9
Trophy meat distribution	933,327	4.9
Use of own game	739,629	3.9
Campsites	520,760	2.7
Shoot and sell	504,883	2.7
Craft sales	161,994	0.9
Interest earned	161,807	0.8
Premium hunting	43,600	0.2
Veld products	39,000	0.2
Miscellaneous	34,788	0.2
Live game sale	0	0
Total	19,045,854	100.0

Source: NACSO (2007:28). Tabularisation: F. Becker 12/2007

of the people and to compensate for the costs of biodiversity conservation' (Vorlaufer, 2007: 26).

It becomes obvious that quantitative data measuring growth reflect neither success nor failure of CBT projects, appropriately. With Goodwin (2000), one needs to acknowledge that CBT in communal conservancies means pro-poor tourism. Subsequently, this means to create opportunities for those who lost their livelihood. In earlier writings, Ashley (1995), Davis (2001), and Palm and Pye (2001) published their deliberations on this issue. Their concern challenges researchers to conduct studies with pro-poor research concepts, methods and techniques that 'focus on finding the niches and framework(s), necessary for the negotiation of poverty alleviation opportunities that likely secure and improve the physical and emotional survival of people' (Becker, 2006b) and their vulnerable communal lands.

The entrepreneurial management request

The answer to the question, how CBT introduces local development and, ultimately, for whom, remains for the evaluation of individual cases. One may emphasise that the CBT approach succeeds more confidently in communal conservancies with management capacity and experience in CBNRM and CBT. With these agents instituted for local progress, conservancy members stand a good chance to push growth and development in the area under their governance. The prerequisite that implies any benefit from igniting tourism resources entails a caring input of means of production such as natural resources including wildlife, culture and human skills like artisanship. Any input of this kind implicates the preparedness to compete the formal tourism sector's claims on conservancies with all means available in the informal sector economy. The indicated competition would rather launch, complement or diversify possible local and regional economic cycles, marking actor-induced spatiality through private entrepreneurship. At their best, such economic cycles constitute the platform for and, simultaneously, the goal of sector and spatial planning. From this point on, CBT in conservancies achieves the property and function of development nodes that qualify for the inclusion in the spatial planning of regional to national territoriality.

Epilogue: Community-based Tourism in the Light of Natural Hazards

In the fall of 2007, Namibia's people, wildlife, domestic tourism destinations and their visitors experienced and even endured extreme events, climatic in origin. Severe drought, sunburnt landscapes and starving wildlife shadowed the visitors' recreational expectations in some parts of the country. In February and March 2008, the long awaited 'good rains' turned the vegetation of the tourism landscapes into open plains and mountains of lush green and, running ephemeral rivers likely tempted self-driving tourists to become adventurous crossing them. While tourists accepted and perhaps enjoyed the unexpected otherness of extreme weather in a perceived arid environment, people in communal areas continued to fight for survival in flooded regions of northern Namibia; having lost homesteads, crops, infrastructure, potable water, security of basic human needs.

It is too early to present an assessment of damage to the material investments of conservancies with CBT, following the natural disaster. Against the background of concepts, strategies and quantitative results believed to define success, the hazard of either flood or drought striking peripheral communal lands under tourism development reminds the sector as well as the spatial planner to review perhaps the 'tool box' of CBT development. This exercise should examine how education and training in disaster prevention could become part of people's capacity. It might enable members of tourism-based conservancies to recognise mismanagement of resources and their development with an informed attitude to change of human behaviour in planning; mitigating, if not avoiding affects on material investments in the event of approaching hazards. The mindfully integration of local communities into the process of establishing classified tourism planning regions appears to be a necessity. Such regions might assist in sustaining any promising endeavour implicit in the development of communal tourism destinations characterised by scarce and fragile resource conditions with disastrous risk probabilities such as flood following drought.

References

Albl, S. (2001) Conservancies auf kommerziellem Farmland in Namibia. Unpublished MA thesis, Diplomarbeit, Universität Trier. Fachbereich VI: Geographie/Geowissenschaften.

Apostolakis, A. (2003) The convergence process in heritage tourism. *Annals of Tourism Research* 30 (4), 795–812.

Ashley, C. (1995) *Tourism, Communities, and the Potential Impacts on Local Incomes and Conservation* (DEA Research Discussion Paper, 10). Windhoek: Ministry of Environment and Tourism.

Barnard, P. (ed) (1998) *Biological Diversity in Namibia: A Country Study*. Windhoek: Namibian National Biodiversity Task Force.

Becker, F. (2002) Tourismus, Landschaftsschutz und lokale Entwicklung – Perspektiven des Wandels in Namibia. *Petermanns Geographische Mitteilungen* 146, 68–75.

Becker, F. (2003) Product design in higher education: The 'tourism' degree of the University of Namibia. In P. Naudé and N. Cloete (eds) *A Tale of Three Countries: Social Sciences Curriculum Transformation in Southern Africa* (pp. 205–220). Lansdowne: Juta.

Becker, F. (2006a) Development of space in independent Namibia: Deliberations on the projection of politico-economic and socio-cultural transformation. In H. Leser (ed.) *The Changing Culture and Nature of Namibia: Case Studies* (pp. 17–35). The Sixth Namibia Workshop, Basel, 2005. In Honour of Dr Hc. Carl Schlettwein (1925–2005). Basel: Baseler Afrika Bibliographien.

Becker, F. (2006b) Institutional reform, urbanisation, politico-economic and socio-cultural transformation in northern Namibia: A pro-poor investigation into poverty alleviation opportunities. Invited Poster. Volkswagen Foundation Workshop on Resources, Livelihood Management Reforms, and Processes of Structural Change. Gobabeb/Namibia, 18–23 September 2006.

Chirawu, T.O. (2004) Land ownership and sustainable development in Namibia. In D. Olowu and R. Mukwena (eds) *Governance in Southern Africa and Beyond. Experiences in Institutional and Public Policy Reform in Developing Countries* (pp. 227–246). Windhoek: Gamsberg Macmillan.

Davis, A. (2001) Communal area conservancies in Namibia – a unique and success-ful initiative. In P. van Schalkwyk (ed.) *Conservation. The Importance of the Environment for Development in Namibia* (special edn, pp. 12–13). Windhoek: Travel News Namibia.

Duncan, J. (1994) Landscape. In R.J. Johnston, D. Gregory and D. M. Smith (eds) *The Dictionary of Human Geography* (3rd edn, pp. 316–317). Oxford, Cambridge, MA: Basil Blackwell.

Fleminger, D. (2006) *The Cradle of Humankind.* Johannesburg: South Publishers.

Goodwin, H. (2000) Pro-poor tourism. Opportunities for sustainable local devel-opment. *Development and Cooperation* 5, 12–14.

Granö, J.G. (1929; 1997) *Pure Geography.* Baltimore. Johns Hopkins University Press.

Herbert, D.T. (1995) *Heritage, Tourism and Society.* Wellington House: Monsell Publishing Ltd.

Hinz, M.O. (2002) Two societies in one – Institutions and social reality of tradi-tional and general law and order. In V. Winterfeldt, T. Fox and P. Mufune (eds) *Namibia, Society, Sociology* (pp. 197–207). Windhoek: University of Namibia Press.

Hüser, K., Besler, H., Blümel, W.D., Heine, K., Leser, H. and Rust, U. (2001) *Namibia. Eine Landschaftskunde in Bildern* (Edition Namibia 5). Göttingen/Windhoek: Klaus Hess.

Long, S.A. (ed.) (2004) *Livelihoods and CBNRM in Namibia: The Findings of the WILD Project.* Final Technical Report of the Wildlife Integration for Livelihood Desertification Project (WILD), prepared for the Directorates of Environmental Affairs and Parks and Wildlife Management. Windhoek: The Ministry of Environment and Tourism, the Government of the Republic of Namibia.

McIvor, C. (2000) In search of the lost paradise. Tourism between illusion and real-ity. *Development and Cooperation* 5, 19–20.

NACSO (2007) *Namibia's Communal Conservancies: A Review of Progress and Challenges in 2006.* Windhoek: NACSO.

National Trust for Historic Preservation (2008) At http://www.culturalheritage-tourism.org. Accessed 12.2.2008

Palm, P. and Pye, O. (2001) *The Potential of Community Based Tourism and the Role of Development Cooperation in Namibia* (Tropical Ecology Support Programme). Eschborn.

Poria, Y., Reichel, A. and Biran, A. (2006) Heritage site perceptions and motivations to visit. *Journal of Travel Research* 44 (3), 318–326.

Prentice, R.C., Witt, S.F. and Hamer, C. (1988) Tourism as experience: The case of heritage parks. *Annals of Tourism Research* 25 (1), 1–24.

Republic of Namibia. Ministry of Environment and Tourism (2001) *Parks and Wildlife Management Act* (pp. 23–24) (Working Document). Windhoek (Unpublished manuscript).

Republic of Namibia. National Planning Commission (2001) *Second National Development Plan* (NDP 2). Windhoek.

Republic of Namibia. National Planning Commission (2007) *Third National Development Plan* (NDP 3). Windhoek.

Seely, M. and Zeidler, J. (2002) Land distribution and sustainable development. In V. Winterfeldt, T. Fox and P. Mufune (eds) *Namibia, Society, Sociology* (pp. 75–84). Windhoek: University of Namibia Press.

Shaw, G. and Williams, A.M. (1994) *Critical Issues in Tourism. A Geographical Perspective*. Oxford: Basil Blackwell.

South African Places (2006) *Cradle of Humankind*. On WWW at http://www.places.co.za/html/sterkfontein_caves.html. Accessed 6.12.2006.

South Africa's World Heritage Sites (2007) *Cradle of Humankind*. On WWW at http://www.sa-venues.com/attractionsga/cradle-of-humankind.htm. Accessed 12.12.2007.

Vorlaufer, K. (2007) Kommunale Conservancies in Namibia: Ansätze der Biodiversitätssicherung und Armutsbekämpfung? *Erdkunde* 61, 26–53.

WTTC (2000) *Tourism Satellite Accounting Research. Estimates and Forecasts for Government and Industry*. TSA Research. Summary and Highlights: Sub-Saharan Africa – Highlights. London.

Chapter 7

Commercialization of National Parks: South Africa's Kruger National Park as an Example

DAVID M. MABUNDA and DEON WILSON

Introduction

Conservation agencies the world over suffer from chronic financial problems that inhibit them from carrying out their conservation mandate adequately (James, 1999). Until the funding noose started tightening, conservation agencies worldwide did not see the need to engage in business/ commercial practices such as profit-making and marketing (Eagles, 1997). There was no attempt to ensure that revenue-earning activities generated surplus income, based on real costs of buildings and maintenance, and operating costs (Hughes, 2003). The result has been an over-dependency on state subsidies without much attention being paid to creating alternative options of revenue generation (Eagles & Wind, 1994). It is believed that many conservation agencies of the world are cash-strapped or survive on shoe-string budgets (Eagles *et al.*, 2001; Van Sickle & Eagles, 1998).

In the 1980s, it appeared that virtually every state in southern Africa was slashing conservation budgets, thus funding became a risk that had to be taken into consideration in future park planning processes (Hughes, 2003). Reynolds (1995) noted that the National Park Service in the United States, the largest government nature tourism provider in the world, faced huge budget cuts in 1995. Figgis (1993) and Wescott (1995) reported that Parks Australia was severely short of funding. In Queensland, for example, the National Parks Agency complained that the recent expansion in the number of national parks was not matched by an increase in funds to meet operational costs (Dickie, 1995). The results of such under-funding manifest themselves in infrastructure in a state of poor repair, some facilities closed during peak holiday periods due to safety risks and insufficient visitor information services (Eagles, 1995).

The question asked in this chapter is whether protected areas can use commercialization as an alternative revenue generation source to achieve financial viability. The outsourcing of tourist accommodation in the rest camps of the Kruger National Park (KNP) is used as a case study to attempt an answer to the question.

Kruger National Park

The KNP is renowned for its unparalleled wildlife management in the African continent, its diversity of animal species and its variety of vegetation zones. It covers an area of 1,948,528 ha (19,455 km²) and lies between 22°25′ and 25°32′ latitude South and between 30°50′ and 32°2′ longitude East. Close on half of the KNP falls within the Limpopo Province and the remaining half in Mpumalanga Province, the western boundary of the park being a rather arbitrary line across the two provinces (Figure 7.1). The Lebombo Mountains form the eastern border between the KNP and Mozambique. The Limpopo River forms its northern boundary with Zimbabwe, while the Crocodile River is its southern limit. In geographical size, the KNP is equivalent to the state of Massachusetts in the United States and Wales in Britain.

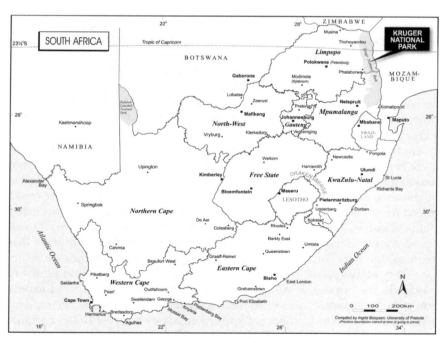

Figure 7.1   Map of South Africa showing the Kruger National Park

Currently, South Africa's National Parks (SANParks) generate 80% of its revenue from accommodation and admission fees in the KNP. This revenue helps to cross-subsidize the other national parks with the exception of Table Mountain, Tsitsikama, Augrabies, Kgalagadi and Addo National Parks, which all generate surplus income after calculating overheads. The remaining 15 national parks and the Groenkloof Head Office (in Pretoria) depend on the KNP for financial survival.

Like the rest of the world conservation institutions, the KNP suffered perennial shortages of money to such an extent that its initial tourism facilities were established with the help of donations. According to Carruthers (1995), this was because charges were minimal and the revenue earned from tourists did not meet the initial high public expectations raised. The government of the day was unwilling to commit financial support for tourism development because it believed that with a facility like the KNP, the National Parks Board (NPB) should have been capable of generating sufficient revenue to attain financial viability (Carruthers, 1995). The perennial shortage of money means that the provision of accommodation usually lagged far behind visitor demands, leading to frequent complaints of overcrowding, lack of facilities and poor conditions of hygiene (Carruthers, 2001).

The KNP staff was unable to manage all aspects of tourism and in 1930 experiments were made with outsourcing to private contractors and the South African Railways and Harbours (SAR&H). The private contractors operated the shops and restaurants, whereas the SAR&H marketed the Park, employed gate officials, sold entry tickets, transported tourists by train to the Park and managed the accommodation facilities on behalf of the NPB (Carruthers, 2001). After a disastrous experience of poor service delivery by the private operators and the SAR&H, the NPB, for the first time, took over the management of these facilities for its own account in 1951 (Struben, 1953).

Traditionally, entrance and overnight accommodation fees in the KNP were kept low in order to facilitate accessibility to what was regarded as a national heritage and as part of a broader social or educational objective. Such an approach might have been appropriate in a country that could afford it, because easy access can cultivate an appreciation for wildlife and political goodwill (Msimang *et al.*, 2003). However, South Africa is a developing country facing many post-apartheid socio-economic problems that exert competing claims on the public purse (Mabunda & Fearnhead, 2003). Only the relatively affluent residents and foreign tourists could afford to pay such 'low' prices. It would be therefore unrealistic to expect the new South African government to subsidize the use of this natural heritage for affluent tourists. Undercharging simply increases the cost to the national treasury of maintaining the Park estate and it represents an opportunity cost to biodiversity (Msimang *et al.*, 2003). According to the McKinsey

Report (2002), low pricing of products and services at all national parks, including the KNP, have contributed to the current financial under-performance and it has therefore become imperative for SANParks to seek alternative ways of managing its facilities in a cost-effective and profitable manner. One such alternative is commercialization as a conservation strategy.

Commercialization as a Conservation Strategy

The term 'commercialization' evokes different emotions in different people because of the wide spectrum of management options it can entail. It can mean the development of a basic service ethic to complete privatization of parks involving the selling of both land and infrastructure. At SANParks, commercialization implies an intention to generate additional revenue as a means of ensuring better conservation of national parks (Fearnhead, 2003). This additional revenue is generated by granting the private sector the opportunity to operate within national parks but without alienating any of the assets (SANParks, 2000a). Private companies are awarded concessions within national parks, that is, an opportunity to build and operate a tourism facility or taking over an existing line of business like the historically under-performing bush/private camps, shops, restaurants, petrol stations, laundry, cleaning, garden services and new concession lodges on virgin sites (luxury lodges with traversing rights to cater for the top-end of the market (IFC, 2001).

Concessionaires are expected to design, construct, operate and maintain assets they take over from SANParks for a contract term ranging from nine to 20 years under strict operational procedure manuals (SANParks, 2000a). This public–private partnership (PPP) must exhibit the correct mix of financial strength, requisite business experience and strong empowerment credentials with a 20% or more shareholding (SANParks, 2000b). It is estimated that at maturity, tax receipts will be R60 million per annum. Commercialization at SANParks is a conservation strategy that will allow staff to concentrate on the core business, that is, conservation of biodiversity, eliminate inefficiency and wastefulness, improve cash flow and service levels and outsource peripheral businesses to the private sector while empowering communities (Fearnhead, 2003).

Resistance to Commercialization

There are many conservationists who do not share the view that national parks and other nature reserves should be financially self-reliant. On the international scene, the Third World Network (TWN) organization recently voiced its opposition to the worldwide privatization and construction of environmentally unfriendly buildings, thus violating the principle of 'touching the earth lightly' in a quest to lure more tourists with dollars. The TWN appealed to the Seventh Conference of Parties (COP 7) at the Convention on Biological Diversity (CBD) meeting in Malaysia from 9 to 27 February 2004, to resist over-commercialization of conservation areas. The TWN argued that in the tourism-related 'PPPs' equation, the 'private' appears to have gained absolute primacy. Often these cooperation agreements result in the weakening of government's decision-making power and community participation, with the domineering private partners eventually appropriating the public agenda (Pleumarom, 2003).

The CBD was requested to improve the outdated 2001 Guidelines on Biodiversity and Tourism and put in place tighter measures that will put an end to the insensitive 'invalorization' of biological diversity and the 'Disneyfication of parks'. According to Pleumarom (2003), former term refers to the concept of calculating the financial equivalent value for certain natural resources that attract paying tourists for financial gain by private investors at the expense of ecology, whereas the latter refers to the over-development of entertainment structures resembling Disneyworld in Florida (USA) within national parks and world heritage sites across the world in quest to make money for survival.

In South Africa, there was a wave of rage and vehement opposition when SANParks first mooted the concept of commercialization (M-Net, 2002). The Kumleben Commission of 1998 concluded that 'nature conservation as such can never be self-supporting. ... It is therefore short-sighted and fallacious to expect a protected area to become economically self-sufficient' (Kumleben *et al.*, 1998). Commercialization faces considerable opposition from the public and such a scenario compounds the complexities faced by parks towards attaining financial viability. Critics of commercialization claim that the approach has a controversial history and is not sustainable (Venter, 2002). They further claim that it leads to exclusivity by hiking tariffs and the wholesale selling of fauna and flora to raise money for investment. It would appear, according to these critics, that commercialization focuses too much on profit-making rather than on biodiversity conservation (Macleod, 2003).

While there is a legitimate need to adopt a commercial and business approach to managing the KNP's products and services, there were fundamental concerns raised by internal and external opponents of commercialization. Internally, there were differences of opinion between conservationists and top management on the direction and extent that commercialization should take. There were also untested accusations of certain Board members peddling the commercialization agenda for their own self-interests. The natural science researchers feel that they were not fully involved in the feasibility studies to determine the impact of this strategy on biodiversity conservation. KNP-based officials feel that they have no control over the relationship between themselves and all the concession operators in the park because they are micro-managed from Pretoria (Personal information, Danie Pienaar, Head of Scientific Services, 13 May 2003).

Externally, there was a big outcry from certain quarters of the public when the shops and restaurants were outsourced in 2001. The complaints ranged from deteriorating standards, high prices and over-commercialization of nature – trends that go against the primary purpose of establishing a national park to preserve nature in its pristine natural state. Fears of both 'invalorization' and the 'Disneyfication' of the KNP and other parks are starting to manifest themselves in the eyes of the public. It is against this backdrop that more studies should be conducted on the acceptability and viability of commercialization as an alternative source of revenue for parks.

Outsourcing of Rest Camp Accommodation in the KNP

In 2001, SANParks alluded that it might consider commercializing its rest camp accommodation product in the KNP and other bigger parks like Addo in future as part of its commercialization strategy in a quest to address the need for increased revenue for conservation purposes (IFC, 2001). It had therefore become imperative to conduct a survey to explore public opinion in this regard given the negative responses experienced after the outsourcing of the shops and restaurants referred to earlier.

The main rationale for this survey was to determine tourists' opinion about the outsourcing of rest camp accommodation in the KNP and whether the public perceives such a move to be acceptable or not and whether they would still support the park if prices were to increase. Other reasons for the study are to determine whether the frequency of visits would increase or decrease if the KNP decided to outsource its accommodation facilities and whether a differential pricing policy for international and local tourists is desirable. In addition, tourists were asked to rate their overall satisfaction with the accommodation facilities as they currently stand and whether there is a need to upgrade and make the facilities luxurious like those offered by the private nature reserves on its borders (Box 7.1).

Box 7.1   Tourist Satisfaction in the Etosha National Park in 2008

JARKKO SAARINEN, SUSANNE SCHOLZ and JULIUS ATLHOPHENG

Etosha National Park is the second largest national park in Namibia. The Park's size is 22,270 km^2 and its main attraction along with the wildlife is the vast Etosha Pan. This case study focuses on the visitor's satisfaction and motivations to visit the park. The data were collected in June–July 2008 in the Etosha National Park rest camps of Namutoni and Halali and it consists of 178 survey questionnaires and 34 interviews (Saarinen *et al.*, 2008). The main motivation for visiting the Park was the wildlife (61%) and the natural features (13%) in general. In addition, social motivations related to spending time with family and/or friends were mentioned by 9% of the respondents. Other reasons mentioned were associated with the 'brand' of Etosha being a top destination, suitable climate at the time (season), for example.

In general, tourists were satisfied with the services in the Park. The main motivation was served, that is, 87% of the visitors were satisfied with the wildlife viewing opportunities in the Park (Table 7.1). Over 80% of the respondents rated the cleanliness of the resort areas as good or very good. Also safety in the Park, experiences at camp waterholes, service at the reception and quality of accommodation were evaluated very positively by most respondents.

The most negatively seen issue was the selection/variety in the camp shops, which was evaluated as poor or very poor by over 21% of the respondents. 'Namibian feel' in the camps and choice of food at camp restaurants were also seen as relatively poor. Still, most of the respondents were rather satisfied with these elements – including the selection in the camp shops – than dissatisfied. However, many of the respondents ranked general skills of the staff, overall service in camps, level of nature protection, service at the reception, quality of accommodation and service at restaurants as average.

Methodology

Data were collected by a two-page questionnaire. The questionnaire was administered in English only, so that foreigners who could not understand English were not included in the sample (and were treated as missing data).

The researcher, assisted by five research assistants (Masters degree students from Unisa: Department of Psychology), administered the questionnaire verbally. This was done during the South African school holiday period covering the week of 26–28 March 2003. Some respondents filled in the questionnaire themselves. Respondents were informed about the motivations for the research and the importance of their opinions was stressed. Respondents were also assured of anonymity and confidentiality.

The convenience sample consisted of 316 tourists to the KNP on the 26, 27 and 28 March 2003 (Table 7.2). After a brief pilot study, the researchers found that the best areas to find respondents were the restaurant/shop/picnic areas during breakfast and lunch hours. People leaving and arriving at these areas were asked to complete the questionnaire. Participation

Table 7.1 Tourists' satisfaction with the services and opportunities in the Etosha National Park (%)

	Very good	Good	Neither good nor poor	Poor	Very poor	No know-ledge	N
Wildlife viewing in the Park	40.5	46.4	6.0	1.2	0.6	5.4	168
Cleanliness of the resort area	34.5	47.6	7.1	1.2	1.2	8.3	168
Safety in the park	32.1	46.1	7.9	0.6	0.0	13.3	165
Experience at camp waterholes	29.8	41.1	9.5	0.6	0.6	18.5	168
Service at the reception	24.0	37.1	13.2	3.6	1.8	20.4	167
Quality of accommodation	21.6	37.7	10.8	0.6	0.0	29.5	167
Upgraded camps after 2006	18.4	16.6	8.6	0.6	0.6	55.2	163
Nature protection in camps	18.0	38.9	15.0	1.2	0.6	26.3	167
Overall services in camps	17.4	52.1	16.8	1.2	0.0	12.6	167
Namibian 'feel' in camps	17.3	32.7	31.5	6.0	2.4	10.1	168
Skills of the staff in camps	15.1	39.2	15.7	1.2	1.2	27.7	166
Service at the restaurant	9.6	26.9	10.8	2.4	1.2	49.1	167
Choice of food at restaurant	5.4	22.1	15.8	3.6	1.8	49.4	166
Selection in the shops	4.2	20.6	27.9	16.4	5.1	24.8	165
Quality of the guided night game drives	3.1	6.8	8.1	0.6	0.0	81.4	161

Note: No knowledge means that the respondent has no experience with the issue asked (N = 161–168).

Table 7.2   Respondents from according to origin

Country	Count	Percentage
South Africa	245	77.5
Africa	2	0.7
Overseas	69	21.8
Total	316	100.0

was voluntary. Most people were friendly and willing to participate. Only one out of every 20 people refused and this was usually a foreigner who could not speak English. Participating tourists were from Berg-en-Dal, Skukuza, Lower Sabie, Pretoriuskop and Satara. People at the picnic spots Afsaal and Tshokwane were also sampled.

Once all the raw data had been coded, the analyses were performed using the analysis of variance method (ANOVA). The data obtained were nominal and ordinal, thus only descriptive analyses of an exploratory nature were done. The frequency and percentage of responses for each satisfaction level were calculated for South African, foreign and mixed samples.

To determine whether the majority of the respondents were in favour of or against outsourcing, the frequency and percentage of responses falling into the For/Unsure/Against outsourcing categories were calculated for all three groups (Figure 7.2). The reasons given by respondents for their opinions about outsourcing were examined and discussed. To determine

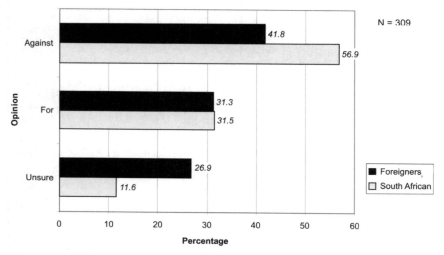

Figure 7.2   Percentage of responses of South Africans and foreigners regarding the outsourcing of accommodation

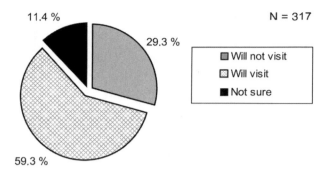

Figure 7.3 Percentage of responses on price increase and continued visits to the park

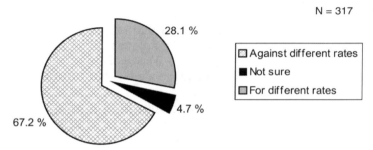

Figure 7.4 Percentage of responses about different accommodation rates for foreigners and South Africans

whether price increases would affect the respondents' frequency of visits to the KNP, total scores and percentages were calculated for the whole sample (Figure 7.3). Qualitative remarks regarding this item were also discussed.

Similarly, to determine overall opinions about whether foreigners should be charged more for accommodation than South Africans, results were calculated for the whole sample and for the South African and foreign samples separately (Figure 7.4). In addition, the responses from the portion of the total sample that were in favour of charging more for foreigners were further examined. The percentage of responses from this group that wanted to charge 50% more/double the price/three times the price/more than three times the price/other, respectively, were calculated (Figure 7.5).

Accommodation Satisfaction

As mentioned, 294 respondents stayed overnight in the park. Four of them did not rate their level of satisfaction with accommodation. Thus, the responses of the remaining 290 were tabulated. The frequency of the

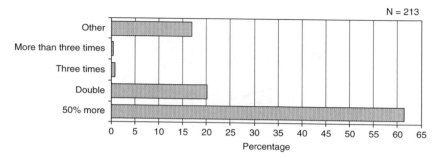

Figure 7.5 Percentage of responses regarding an appropriate rate of increase for foreigners

response 'Very satisfied' implies that the majority of the sample was very satisfied with the accommodation (Figure 7.6).

Out of the sample of 290 respondents that rated the accommodation, 231 were South Africans and 58 were foreigners (Figure 7.7). The most frequent response for both samples was 'Very satisfied'. Only 6.1% of the South Africans and 6.9% of the foreigners were either 'Not at all satisfied' or 'A little dissatisfied'.

Opinions about accommodation outsourcing were explored for the total sample as well as for the South African and foreign samples separately. In all three groups, the largest percentage of the samples was found to be opposed to the outsourcing of accommodation to private operators. Eight respondents did not answer the question. Four of these were

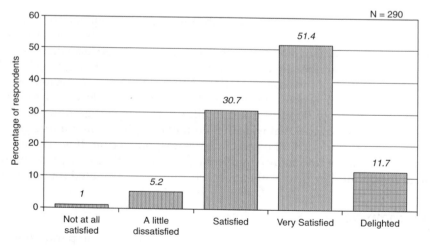

Figure 7.6 Percentage of responses to different levels of satisfaction with accommodation

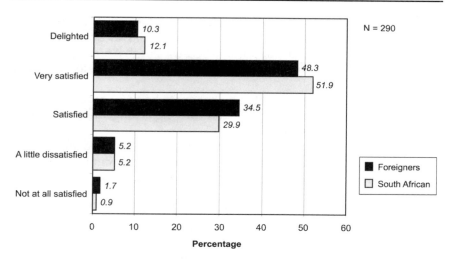

Figure 7.7 Percentage of responses to different levels of satisfaction with accommodation of the South African and foreign samples

South Africans and four were foreigners. They were thus excluded from the results. The one respondent who did not indicate his/her country of origin was included in the results for the total sample but not in the other two samples (see Figure 7.2).

Respondents who were opposed to outsourcing gave a number of different reasons for their opinions. These are as follows:

- The money paid for accommodation must be used for conservation and not for the enrichment of private operators on a public asset built with the tax-payers money.
- Will the park still earn sufficient money to run its operational functions if it depends on the percentage of turnover cuts?
- The park must conserve its culture and must not become commercialized ('Disneyfication').
- Outsourcing will result in an increase in prices, making the park inaccessible for South Africans. It will no longer be a 'national' park.
- Tourists come to the park for peace and tranquility and to see the animals, not for five-star accommodations.
- Accommodation can be upgraded without outsourcing. Better supervision and management are needed.
- The restaurants and shops are worse now since they have been outsourced. They are too expensive, commercialized and service and quality have deteriorated. The same might happen with accommodation.
- The present staff's jobs could be threatened and the local community will not benefit.

- Pensioners are already struggling to afford the park. What will happen if prices were increased? There are no special rates for pensioners (*note*: which is not fully applicable because South African pensioners enjoy a 40% discount rate applicable to periods outside the public school holidays).
- Bookings will be complicated by having to book for accommodation separately from park entrance and catering arrangements for larger groups will also be complicated.
- There will be confusion of responsibilities regarding the role of SANParks and the role of private operators. For example, who will take care of lighting, gardens and roads in the camps?

Respondents who were unsure raised the following concerns:

- It depends on who the operators are. If it ends up like the restaurants and shops then they do not want it.
- It depends on whether there will actually be improvements in service and accommodation, the restaurants are still using utensils and furniture inherited from SANParks yet prices have more than doubled.
- It depends on whether the local community will benefit or not.
- It depends on price increases, prices have more than doubled without improvement on quality of products.
- It depends on whether the present staff will be negatively affected, many staff members who were passed on to the new operators have since disappeared.

Respondents who were in favour of outsourcing of accommodation gave the following reasons:

- outsourcing will be good because it will create competition; and
- if accommodation improves there will be more tourists.

The majority of the respondents showed concern over the possible rate of increase of the price of accommodation, and said their continued visits to the park would depend on the level of price increases. Many respondents, who said that they would still visit the park, stated that they would visit less often, for example, once a year as opposed to three times a year. Often they added that they would camp instead of staying in chalets and other types of brick and mortar accommodation. Respondents who said that they would no longer visit the park mentioned that they would rather go to other more affordable parks (see Figure 7.3).

Different Accommodation Rates for Foreigners

The majority of the sample, 67.2%, agreed that there should be different rates for accommodation for foreigners and South Africans. Most respondents commented that this would be fair since South Africans are already paying tax towards maintaining the conservation estate. Other responses were that this is done in many other parts of the world already and that most foreigners from countries with stronger currencies could afford to pay more. It would not be fair to expect taxpayers in a developing country to subsidize rich tourists from foreign countries.

Respondents who disagreed with the idea of differential rates were concerned about a possible negative impact on tourism and South Africa's attractiveness as a destination in the highly global and competitive tourism market (see Figure 7.4).

The 213 respondents, who agreed that foreigners should pay different rates, were asked to indicate what kind of difference they would consider to be fair. The majority of this group considered '50% more' to be fair (see Figure 7.5). Very few respondents considered 'three times the price or more' to be realistic or fair. A substantial portion (labelled 'other') did not agree with the given options and responded with comments like 'Prices must be related to the exchange rate' or 'Foreigners must pay in dollars'. Some of the respondents from the 'other' group felt that rates of between 15% and 30% higher would be fair, but not more.

When comparing the opinions of South Africans and foreigners about different rates, it was found that a higher percentage of foreigners (39.4%) were opposed to different rates compared to the South African rating of 24.9%. However, the majority of both samples agreed that there should be different rates for foreigners (71.4% of South Africans and 53.5% of foreigners) (Figure 7.8).

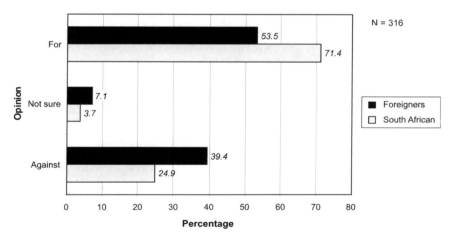

Figure 7.8 Responses of South Africans and foreigners regarding different accommodation rates for foreigners. The respondent who did not specify country of origin is not included in the results below

Conclusions: Discussion and Recommendations

The primary objective of the survey was to determine what tourists to the KNP thought about the outsourcing of accommodation. Clearly, for this particular sample, the majority of tourists does not want accommodation to be outsourced and are satisfied with the accommodation as it is. There is a fear regarding price increases and a loss of the 'Kruger Culture' through 'Disneyfication' if outsourcing were to occur. Many people are unhappy about what they perceive to have happened with the restaurants and shops regarding commercialization, price increases, poor quality and service, and foresee the same trend affecting accommodation if outsourcing were to be introduced.

Concern was raised about the impact that outsourcing would have on the local communities and present staff. The researcher recommends that if outsourcing of accommodation is to be implemented, a safety net needs to be devised to protect staff from the possible future shedding of jobs by private operators motivated by profit-making ambitions. It is further recommended that favourable entry fees and overnight packages be offered to local communities.

With respect to the implementation of differential accommodation rates for foreigners and locals, the majority of the South African and foreign respondents supported the idea. However, most respondents agreed that an increase of 50% for foreigners or less would be acceptable. The researchers recommend that, if differential pricing is implemented, the level of increase should be determined with caution. A number of respondents mentioned that there were many other private game parks that tourists could visit if price increases were to be ridiculously high. There was concern about foreign tourists choosing to visit other natural destinations such as provincial reserves and national parks in other African countries rather than in the KNP. Prices related to the exchange rate or dollar rates should be explored within the ambit of Treasury rules.

It is recommended that, if SANParks would still seriously consider outsourcing its rest camp accommodation product in future, a more comprehensive study be conducted using the present study as a pilot. In addition, if outsourcing were to occur, strict regulations and contract management capacity (Service Level Agreement) should be implemented regarding the preservation of the 'Kruger Culture'. Caution would also have to be exercised regarding possible resultant price increases. Although most people responded by saying that they would still visit the park, many of this group added that they would visit less often and come as day visitors while staying at cheaper but comfortable hotels outside the KNP or use camping facilities. Thus, it appears that tourism income in the park would be negatively impacted if outsourcing is introduced without due consideration. The consequences of commercialization of accommodation facilities in the KNP would need to be explored extensively in terms of its impact in the wider domestic and international markets in the context of the total destination marketing framework.

The search for lasting and reliable sources of revenue to manage protected areas adequately and effectively is a critical management function as long as the near-universal under-investment in conservation management persists (Wells, 1997). The national demands on the public purse are forcing elected legislators to channel scarce financial resources to education, social services, meeting housing needs, fighting crime, health care and the general reduction of poverty. Conservation ranks very low on the national priority agenda of many governments including South Africa. It is unlikely that the SANParks budget will be substantially increased in the near future. The scenario of perennial shortages of money in the KNP justifies efforts directed at finding innovative alternatives of revenue generation such as commercialization of non-core functions as argued in this chapter.

Commercialization, however, should be a well thought-out strategy focused on functions where a protected area lacks expertise and innovation. If there is anything that the KNP has done well over the years, it is the designing and management of its rest camp product in addition to biodiversity management. No other commercial hotel or hospitality group provides this experience at such affordable budget prices. In the feasibility exercise to test the waters in 2001, SANParks put out feelers to the private sector to test market interest in taking over the management of its accommodation estate. The general response by the private sector was that they would not be in a position to offer the same product within the current cost structure for the same prices (SANParks, 2001).

When analysing the revenue streams of the KNP, it is evident that accommodation generates 80% of the park's income. It would therefore require a five-fold increase in unit or per person pricing to match current revenue derived from accommodation and the rest camp concessionaire would have to pay not less than 50% of total earnings to ensure that the KNP receives reasonable but still inadequate funding to carry out park management functions. Such a step would make the park more inaccessible and unaffordable to the general public. No sensible operator would pay 50% concession fees on total revenue earnings to operate tourism facilities in a national park. The cost of doing such business would be extremely high and unaffordable. The majority of tourists, as argued in this chapter, want reasonable and affordable rest camp accommodation with its rustic character and ambiance and not up-market luxuries. This is in line with the park's public mandate of being accessible to the general public.

Although differential pricing is a worldwide practice, its implementation should be carefully managed to avoid feelings of discrimination on the side of international tourists. The logistics of passport checking to determine nationality would in themselves be cumbersome for both the

Park and tourists alike. It is also difficult to justify two different rates for the same bed or service. Given these challenges, the initiative by the SANParks Tourism and Marketing Directorate to review the Wild Card in addressing the issues raised in this chapter is to be welcomed.

It is suggested that, should the outsourcing of rest camp accommodation still be considered an option, the present survey be extended into an extensive investigation to include respondents outside the park (at shopping malls, airports or their homes) on a random basis. It could be argued that the participants in this survey already had a pre-conceived position about the park's services because they were interviewed while on vacation in the KNP. Nonetheless, the reasonable concerns of the public on this subject should be noted and addressed seriously.

References

Carruthers, J. (1995) *The Kruger National Park: A Social and Political History.* Pietermaritzburg: University of Natal Press.

Carruthers, J. (2001) *Wildlife & Welfare: The Life of James Stevenson–Hamilton.* Pietermaritzburg: University of Natal Press.

Dickie, P. (1995) Money squeeze on parks. *Brisbane Sunday Mail* (May 7). On WWW at http://www.sciencedirect.com.sience.

Eagles, P.F.J. and Wind, E. (1994) Canadian ecotours in 1992: A content analysis of advertising. *Journal of Applied Recreational Research* 19 (1), 67–87.

Eagles, P.F.J. (1995) Understanding the market for sustainable tourism. In S.F. McCool and A.E. Watson (eds) *Linking Tourism, the Environment and Sustainability: Proceedings of the Annual Meeting of the National Recreation and Parks Association.* Ogden: UT: USA Department of Agriculture, Forest Service, Intermountain Research Station.

Eagles, P.F.J. (1997) International ecotourism management: Using Australia and Africa as case studies. Paper prepared for the IUCN World Commission on Protected Areas. Protected Areas in the 21st Century: From Islands to Networks, Albany, Australia.

Eagles, P.F.J., Bowman, M.E. and Chang-Hung Tao, T. (2001) *Guidelines for Tourism in Parks and Protected Areas of East Asia.* Switzerland and Cambridge: IUCN, Gland.

Fearnhead, P. (2003) *The South African National Parks Commercialization Strategy,* SANParks Update Report to the Directorate. Unpublished document, Pretoria.

Figgis, P.J. (1993) Ecotourism: Special interests or major direction? *Habitat International* 21 (1), 8–11.

Hughes, G.R. (2003) Commercialization in the Natal Parks Board: The road to survival. Unpublished manuscript, KwaZulu-Natal Conservation Trust.

International Finance Corporation (IFC) (2001) SANP commercialization programme: Strategy on outsourcing of shops and restaurants. Unpublished manuscript, Pretoria.

James, A. (1999) Institutional constraints to protected area funding. *International Journal for Protected Area Managers* 9 (2), 15–26.

Kumleben, M.E., Sangweni, S.S. and Ledger, J.A. (1998) *Board of Investigation into the Institutional Arrangements for Nature Conservation in South Africa.* Pretoria: Government of South Africa.

Mabunda, D. and Fearnhead, P. (2003) Towards sustainability. In A. Hall-Martin and S. Carruthers (eds) *South African National Parks: A Celebration Commemorating the Fifth World Parks Congress*. Singapore: Horst Klemm.

Macleod, F. (2003) Makro King in parks rescue plan. *Mail & Guardian*, 6 June, 15, Johannesburg.

Mckinsey report (2002) Final meeting. South African National Parks. Unpublished manuscript, Pretoria.

M-Net (2002) 'Privatization of National Parks'. Carte Blanche Programme, Johannesburg.

Msimang, M., Magome, H. and Carruthers, J. (2003) Transforming South African National Parks. In A. Hall-Martin and J. Carruthers (eds) *South African National Parks: A Celebration Commemorating the Fifth World Parks Congress*. Singapore: Horst Klemm.

Pleumarom, A. (2003) Our world is not for sale! The disturbing implications of privatization in the tourism trade. *Paper presented at the International Seminar on "Tourism: Unfair Practices – Equitable Options"*, 8th–9th December 2003, hosted by The Network for Sustainable Tourism Development (DANTE). Hanover, Germany.

Reynolds, J. (1995) National Parks and protected areas in the United States. Paper given to the North American Regional Meeting of the Commission on National Parks and Protected Areas. Banff National Park: Alberta.

Saarinen, J., Scholz, S. and Atlhopheng, J. (2008) *Tourist Satisfaction in the Etosha National Park: Summary*. Oulu: Department of Geography, University of Oulu.

SANParks (2000a) Commercialization process – Technical evaluation of bids report: Recommendations to the Chairpersons' Committee. Unpublished manuscript, Pretoria, 19th November.

SANParks (2000b) Summary of recommendations for the SANP commercialization. Board Meeting Agenda. Unpublished manuscript, Pretoria, 29 March.

SANParks (2001) Commercialization programme: Structuring and market review report. Prepared by the IFC with assistance from White and Case, Richard Davies and Associates, Deloitte and Touche, Business Map, Consulta Research and Albert de Vos (IDC). Unpublished manuscript, Pretoria.

Struben, F.E.B. (1953) A history of the Kruger National Park. *African Wild Life* 3 (7), 209–228.

Van Sickle, K. & Eagles, P.F.J. (1998) Budgets, pricing policies and user fees in Canadian Park's tourism. *Tourism Management* 19 (3), 225–235.

Venter, O. (2002) Milking the conservation cow: KNP privatization. *50/50 television programme*. South African Broadcasting Corporation: Johannesburg.

Wells, M.P. (1997) *Economic Perspectives on Nature Tourism, Conservation and Development*. Washington DC: World Bank.

Wescott, G. (1995) Victoria's national park system: Can the transition from quantity of parks to quality of management be successful? *Australian Journal of Environmental Management* 4 (2), 210–223.

Chapter 8

Natural Resource-based Tourism and Wildlife Policies in Botswana

JULIUS ATLHOPHENG and KUTLWANO MULALE

Introduction

Botswana is a semi-arid country with a total land area of 582,000 km^2 and a population density of 2.7 per km^2. It is characterized by periodic droughts that partly explain the country's fragile environment. Like most other developing countries of the world, Botswana relies upon natural resources for development and general livelihoods. Rural livelihoods especially depend upon natural resources such as water, forests, rangeland and so on, which hinge strongly on climatic factors. Likewise, the tourism industry in Botswana is dependent on natural resources as well, which are shared with other land uses. This translates to multiple uses of scarce and fragile natural resources of the sensitive semi-arid environment. The key to the management of natural resources in Botswana is the rationalization of land use and management. Botswana has a land tenure system characterized by communal (tribal/customary), state and freehold systems. The various land uses within this land tenure system are for agriculture, national parks and game reserves, aquatic areas, forest reserves and urban-industrial land. The various land uses, as illustrated in Figure 8.1, point to the amount of land that may host tourism resources or of diverse tourist interest.

Botswana is endowed with bountiful and diverse wildlife resources with the potential to contribute to the growth and diversification of the economy. Botswana's National Development Plan 9 (NDP9) recognizes wildlife as one of the main valuable natural resources together with minerals and rangelands (Ministry of Finance and Development Planning (MFDP, 2003), and as the principal tourism attraction. Wildlife resources in national parks and game reserves are for non-consumptive uses and the wildlife in wildlife management areas (WMAs) are for both non-consumptive and consumptive uses. In all, Botswana's wildlife resources occupy 37% of

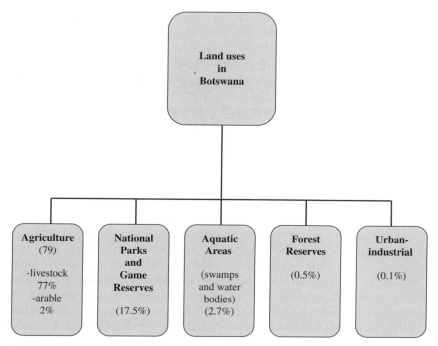

Figure 8.1 Various land uses and their proportions in Botswana (Chanda *et al.*, 2001)

the total land area, of which 17% of the land area is national parks and game reserves, and 20% is WMAs. According to NDP9, the benefits from wildlife include the creation of economic opportunities and diversification with potential for sustainable development, enhancement of environmental stability, contribution to tourism development and provision of aesthetic, scientific, nutritional and educational values (MFDP, 2003). As such sound wildlife management practices are pertinent for the continued existence of the resource base. Considering the importance of the wildlife resource, the Government of Botswana has come up with a number of policies and legislation guiding and regulating the management and utilization of the resource.

This chapter presents some of the relevant policy documents and legislation pertaining to wildlife management and utilization of wildlife in tourism, and note down other competing land uses and their potential threat to wildlife tourism. The chapter also discusses landforms as natural tourism resources in Botswana. The country's terrain is normally characterized as flat to gently undulating. The landscapes found in the country are the flat Kalahari Desert with its various sand dune types and pans: the hilly terrain that dominates the eastern fringe, with occasional isolated hills known as *inselbergs* (kopjes), and some sedimentary ridges that belong to the Waterberg Group. Perhaps the most known feature of Botswana, other than the Kalahari, is the Okavango Delta, which is to the north of the country.

Different landforms provide diverse possibilities for nature-based tourism, ecotourism and a specific form of natural resource-based tourism called geotourism. The international literature on tourism continues to be loaded with emerging terminology. As Maher (2007) has noted 'definitions in tourism are running rampant' ... and that 'new terms are created on a case-by-case basis.' The ever increasing terminology has caused problems for the conceptual base of tourism research. Geotourism, for example, brought conflict between geographical and geological tourism (Buckley, 2006). As a concept, geotourism is a term with dual meanings – either as a mixture of geography and tourism or as a mixture of geology/geomorphology and tourism (Maher, 2007). However, it is still further questionable whether this geotourism is a new form of ecotourism, involving conservation and community aspects. However, as Mehmetoglu (2007) has indicated, based on his case study, not everyone visiting a nature-based attraction is purely interested in nature-related activities (only 1/5 rated relaxing and nature-based activities highly). However, in Botswana, the drive seems to be more towards the nature and viewing of wildlife, which is also highlighted in Botswana National Ecotourism Strategy (BNES, 2002). This may leave cultural and community aspects without any emphasis.

Wildlife Policies and Legislation

Wildlife Conservation Policy of 1986

The Wildlife Conservation Policy of 1986 converted stretches of land that were formerly designated as 'reserved' under the Tribal Grazing Land Policy (TGLP) of 1975 into WMAs. According to NDP9 by the MFDP (2003: 246), the WMAs were instituted by the Department of Wildlife and National Parks (DWNP) and the Division of Land Use Planning to function as migratory passageways for wildlife between national parks and game reserves as they permitted for movement that is necessary for the survival of Botswana's wildlife in the arid environment. The WMAs were further sub-divided by the DWNP and the Division of Land Use Planning into Controlled Hunting Areas (CHAs). These CHAs were subsequently earmarked for various kinds of management and utilization with wildlife management, utilization being the dominant land use. In all, the WMAs were sub-divided into 163 CHAs and are located in the western and northern parts of Botswana. It is within the WMAs that the Government of Botswana granted neighbouring communities the right to use wildlife resources within a CHA subject to government regulations under the community-based natural resource management (CBNRM) programme (see also Ramutsindela, Chapter 10 in this book).

The Wildlife Conservation Policy of 1986 guided implementation of the CBNRM programme in Botswana. The policy holds that the CBNRM community-based organizations (CBOs) ought to economically benefit from CBNRM. The expectation is that, such economic benefit for CBNRM/ CBOs will lead to their supporting state wildlife conservation efforts. The policy aims at regulating the commercial and subsistence use of wildlife resources. DWNP regulates the utilization of wildlife resources by awarding annual wildlife off-take quotas. The Government of Botswana thus issues usufruct rights to local community groups and retain ownership and management responsibilities over the wildlife resources.

Wildlife Conservation and National Parks Act of 1992

The Wildlife Conservation and National Parks Act (Act No. 28 of 1992) resulted in the creation of national parks, game reserves and WMAs in which wildlife conservation and use is the principal land use. The act brought every wild animal, including insects under the jurisdiction of the DWNP. DWNP regulates the utilization of wildlife resources by giving hunting licenses or permits to individuals and wildlife resource use leases to local communities involved in CBNRM for commercial and subsistence utilization of wildlife resources. The Government of Botswana believes that state ownership and regulation of wildlife resource use is essential for sustainable wildlife management and use. However, there are emerging paradigms that see the need to devolve wildlife management to local community institutions captioned under the 'catch phrase', CBNRM.

CBNRM Policy of 2007

Botswana's CBNRM Policy was endorsed by Parliament in 2007 after the CBNRM programme has been running for over 10 years without a policy. The CBNRM Policy allows communities to obtain a 15-year CBNRM lease from the relevant land authority for the commercial use of natural resources. The 'Head Lease' is subject to an approved land-use management plan for the area specified in the lease, an annual land rental payable to the land authority and a resource utilization royalty payable to the Ministry of Environment, Wildlife and Tourism. Communities may receive benefits from the use of natural resources in the area specified in the lease. Communities may sublease or otherwise transfer any commercial natural resource use rights to one or more joint venture partners with prior written permission of the land authority. Where financial benefits are derived from the sale of natural resource concessions or hunting quotas related to particular communities, a portion (65%) of such financial benefits shall be paid into a National Environmental Fund held by the Ministry of Environment, Wildlife and Tourism.

Over the years, CBNRM has been closely associated with wildlife resource utilization, hence it is seen in many quarters as a wildlife thing. The rationale for CBNRM in Botswana is that communities who bear a cost for living close to the natural environment, wildlife resources in particular, have to benefit directly from utilization of the resource in WMAs. Arntzen (2003) noted that while the direct use value is a significant constituent of resource utilization and management, it is imperative to understand that the value of natural resources exceeds the direct use value. Arntzen (2003) outlines other wildlife use values as follows:

(1)   The *indirect use values* do not benefit people directly, but refer to key ecological functions of wildlife.
(2)   The *option value* refers to potential future resource uses and the perceived value of preserving the resource for these. A decline in wildlife resources may limit future use options.
(3)   The *existence value* refers to the perceived value of the mere existence of wildlife irrespective of their use by humans.

Game Ranching Policy for Botswana 2002

The policy is intended at developing a game ranching industry that will offer a commercially feasible but sustainable option to livestock enterprises. The policy deals with game ranching and not game farming. The policy defines game farming as the managed, intensive production of semi- or totally domesticated animal species in small fenced camps or farms under controlled conditions. The policy defines game ranching as the managed, extensive production of free-living wildlife on land fenced in accordance with the fence specification in respect of a given species. Compared to other countries in the region, such as South Africa and Namibia, game ranching in Botswana is fairly nascent, but steadily growing. The Game Ranching Policy expresses the Government of Botswana's desire to further game ranching on freehold, state and tribal land outside National Parks, Game Reserves and WMAs (MTIWT, 2002).

Policies and Legislation Regarding Management of Landforms

Botswana has adopted ecotourism to deal with the management of natural resources in tourism. This BNES (2002) has adopted The International Ecotourism Society definition, that: 'Ecotourism is responsible travel to natural areas that conserves the environment and sustains the well-being of local people.' For the Botswana situation, ecotourism refers to the cultural and natural heritage resources, which may include landscapes and landforms. Thus the tourism industry, based on the ecotourism principles, is meant to be sustainable – economically, culturally, environmentally, and in maintaining tourists' interests.

Some of the major landforms in Botswana, which are of touristic interest, include the Tsodilo Hills and the Okavango Delta in the north, Matsieng ancient footprints site and Kobokwe (Logaga-la-ga-Kobokwe) caves around Gaborone and the Kgalagadi (Kalahari landscape) in the west of the country (Table 8.1). The tourism attractions in Botswana are mainly to the north, around the Okavango Delta. Most tourists visit the country to see wildlife, with wilderness being secondary, according to visitor numbers (BNES, 2002; see also Mbaiwa & Darkoh, Chapter 12 in this book). Thus the ecotourism strategy has wildlife and its habitat as the main foci of sustaining tourism in the country. As reflected by the visitor numbers, rocks as heritage do not feature predominantly; neither is the wilderness or the landforms of Botswana. It is therefore, noteworthy that, the Botswana tourism strategy has not explicitly regarded landforms as a tourism entity that requires a separate tourism policy for the management of these resources. However, Alpizar (2006) noted that, natural environments like landscapes of outstanding beauty may be preserved or set aside to provide recreational or scientific studies.

Some challenges related to the running of tourism in the national parks and other such areas is the lack of financial self-sufficiency, which is a dilemma for developing country governments, as to the economic viability of such parks. The land dedicated to conservation (Alpizar, 2006) is at times about subsidising recreation for visitors from rich countries. In addition, there are negative impacts to the environment and the host communities. Landscapes and the ecosystem resources are multi-functional resources to the communities; however, modern trends of economic monetary returns convert these to simple mono-functional entities (private), which are cushioned by tax incentives and subsidies (Mowforth & Munt, 1998).

Although this kind of transformation processes may be beneficial in the short term (taxes/subsidies), the longer term perspective is that of unsustainable practices as erosion of natural resources ensues, as well as economic inefficiency may outweigh the benefits. Some examples of impacts (direct or indirect) consist of (BNES, 2003) disturbance of wildlife, effects on vegetation such as trampling, fires, introduction of exotic species and litter and vandalism. This may include clearing of bush for roads and other forms of infrastructure. This change in land use is also associated with burdening costs to communities (and not private entities) in the form of non-marketed externalities (de Groot, 2006), a challenge to current and future generations' sustainability. If the user fees are small, especially in developing countries, then this essentially leads to subsidizing recreation for the tourists and visitors from developed countries (Alpizar, 2006). The fees are at times used to protect the environment – reduce excessive

Table 8.1 Landforms and their significance in Botswana

Land division	Main types of landforms found	Major land-use type	Major touristic interests
Sandveld (mainly Kalahari)	Undulating Kalahari Plain with vegetated dunes; active dunes SW of country; calcrete rimmed pans; fossil valleys	Wildlife areas; home to various peoples; cattle grazing; pans as watering points; fossil valleys for winter grazing by wildlife; veldt products; mines	Scenic and unique landscape; wilderness; academic studies; wildlife; people of the Kalahari; recreational activities (e.g. trans-Kalahari road race
Hardveld	Pediments, escarpments with occasional hills (kopjes) and undulating plains. Seasonal rivers; eroded valleys; some alluvium. Few hill ranges and fossil valleys punctuate the landscape	Main population centre of the country in the east – towns and villages; various forms of agricultural practice; mines; roads and rail networks	Urban and rural livelihoods; nature reserves; travel; landscapes; cultural sites; academic studies; recreation; modern accommodation facilities
Lacustrine system	Mainly Makgadikgadi pans – major lake and depression deposits; flat palaeo-lagoon with associated shorelines and major pans; Boteti River	Wildlife parks; cattle farms and ranches; recreational activities; molapo/river arable farming; settlements; soda ash mine	Viewing of wildlife; major pans; wilderness; recreational activities
Alluvial system (mainly the Okavango system)	Mainly alluvial deposits – flat to almost flat delta and river floodplain and swamps; flat fan and sand ridges	Fishing; wildlife; veldt products; various farming/ agriculture practices; tourism	Main tourism centre of Botswana (ecotourism); wildlife safaris; delta and swamp views; flying safaris; diverse accommodation; basketry; cultural sites

impacts on the natural environment by limiting visitors or even use the fees to spread out visitations across the year, thus promoting conservation. Such practices have led to terms like ecotourism being proposed, to differentiate this kind of tourism from mass tourism (high visitation, low user fees, thus more environmental impacts).

Policies and regulations relating to landscapes and landforms as tourism resources are implied in the BNES (2003), where the sustainable landscape use and maintenance could be applied. For the most part, policies and regulations that are in place mainly relate to natural resources management that include agricultural resources conservation, water resources, forest reserves, fish protection and waste management.

Most landscapes are interpreted in terms of the biodiversity issues, that is, what fauna and flora they may host. The Botswana Tourism Board Act (2003) was set to 'market and promote Botswana's tourist attractions, and to encourage and facilitate travel, by local and foreign tourists, to the said destinations'. In this act, there is encouragement for the sector to diversify and introduce other forms of tourism such as cultural, heritage, ecotourism, entertainment, recreational and leisure tourism, supported by an effective marketing strategy. For Botswana, some of the cultural heritage sites, such as the Hills of Kobokwe and Tsodilo, may have diverse cultural meanings, including as ancient rock paintings, caves, traditional and religious belief systems associated with them.

The management of cultural heritage sites is covered in the Government of Botswana (2001). The areas declared as monuments, would be protected and preserved; there would be no development within one kilometre of them, and if development is undertaken, it shall not be incompatible with the preservation of such a feature. There is a requirement for full archaeological and environmental impact assessments, for any activity that will physically disturb the earth's surface. The protected heritage area may be an area of land or region of national value whose authenticity, integrity, surrounding setting or atmosphere are being preserved; structure or building of national value erected before 1 June 1902. It could also be a drawing, painting, carving, ornament, implement, stone tool, bone, pottery or any artefact of national value (Government of Botswana, 2001). No alterations, damage or removal without the Minister's permission, is allowed, and archaeological research in such sites needs permission as well. Thus policies and regulations relating to landscapes and landforms are covered under the Act.

Human–Wildlife Land-use Conflict: A Potential Threat to Tourism

Literature points to escalating trends in human–wildlife conflicts which have become a significant issue in conservation and land management.

The human–wildlife conflict has potential to negatively affect the future availability and diversity of wildlife populations. Botswana's NDP9 acknowledges human–wildlife conflicts to be a major source of controversy between those who suffer damage and the Government (MFDP, 2003). One such controversy cited in the plan is a result of damage caused by wildlife to livestock, crops and other properties, and the inadequate or lack of compensation for such damage.

The designation of land under TGLP as WMAs brought about constraints to agriculture. Within WMAs, agricultural development is restricted as the dominant land use in wildlife management and utilization. Arntzen (2003) noted that the WMAs have a profound impact on the livelihood opportunities of the local population. However, WMAs are in agricultural resource poor areas but rich in wildlife resources. Based on this, Cooke (1985) has argued that the Kalahari sandveld is best suited for wildlife than livestock. However, wildlife resources unlike livestock, belongs to and its use regulated by the state. As such, communities living in settlements within or bordering WMAs perceive livestock rearing as an important source of livelihood as they have direct control over it as compared to wildlife. As such, in instances of livestock–wildlife conflicts, local communities deem wildlife as a menace to their livelihood.

NDP9 acknowledges that the root cause of the conflict is the fact that livestock areas have infringed into wildlife areas. Cooke (1985) documents how the livestock industry expanded into the Kalahari sandveld with the introduction of the borehole technology and the acquisition of a lucrative beef market under the Lome Convention. Subsequent government policies like the first Livestock Development Project (LDP1) of 1972 and the TGLP of 1975 helped open up the Kalahari sandveld for livestock ranching (Cooke, 1985). Cooke (1985) noted that ever since, the Kalahari wildlife populations have been shrinking. Albertson (1998) also documented shrinking wildlife populations in the Okavango region of Botswana as a result of the erection of veterinary disease-control fences. Veterinary fences have been erected to combat the spread of diseases from wildlife to livestock. These fences curtail wildlife migration routes, fence off vital wildlife habitat and block cross-border migrations (Albertson, 1998).

Utilization and Management of Landforms as Tourism Resources

Landscapes and landforms in tourism have diverse meanings, and may be used for recreation and business. It is thus of interest to consider the 'emotional meanings that people attach to environments' (Waitt *et al.*, 2003: 526), while Mehmetoglu (2007: 652) points to 'experiences that are dependent on nature, experiences that are enhanced by nature and experiences for which nature is incidental.' In addition, landscapes are sites of scientific investigation, as the mode of their formation, hence their sustainability are also key. Crofts (1975) analysed the various preference techniques (scenic quality, attractiveness or aesthetics) used in landscape evaluation and noted their subjectivity. These beauty judgments may depend on the individual surveyor's experience, culture, the discipline or training background. Thus, scenic quality assessments may benefit from (improved) randomly selected individuals and groups. Deng *et al.* (2002) indicated attractiveness of parks as enhanced by tourism resources, tourist facilities, accessibility, local communities and peripheral attractions. According to Waitt *et al.* (2003), there are malleable boundaries that define nature, such that touristic experiences may not differentiate between large tracts of irrigated agriculture from physical and biological processes. That 'landscapes become attractions through the meanings that are ascribed to them by promotional agencies and tourists' (Waitt *et al.*, 2003: 524), thus tourists being mere dependents of the tourism brochures. This may have challenges for the validity of nature-based tourism – a wilderness trail (of emotional values and meanings ascribed to landscapes) or ecotouristic responsibilities for both visitor and local communities (where social, economic and environmental benefits accrue to all) or a combination of all the above.

It is noted that, resources may directly or indirectly satisfy human needs. In addition, a resource need not only be seen in terms of human utilization – that the intrinsic values of resources are realized. For instance, climate regulation, soil formation, plants pollination and energy transfer are some of the intrinsic functions that do not need any human utilization aspects – they are normal functions of landscapes and ecosystems. Tourism as a field of study takes on board both the human and non-human values, for example, economic or business ventures that may arise out of tourist operations; as well as promoting conservation of species, landscapes and habitats. Some of the sites to be preserved and/or conserved may relate to spiritual, cultural, scientific and genetic resources. Thus the religious, heritage, historic, medicinal and use of nature for scientific research are some of the natural values of ecosystems and landscapes (Table 8.2).

The land resources policies in Botswana operate to sustain the natural resources. Thus the exploitation is aimed at promoting resource use in an environmentally benign manner. For instance, Saarinen (2006) has noted some of the impacts of tourism to have led to various actual and potential problems that may be social, political, environmental, cultural and economic for destinations and systems. This, he argues, create a need for alternative and friendlier practices (environmentally and to the hosts) in decision-making processes in development, planning and the policies (Box 8.1).

Table 8.2 Summary of some major landscapes/landforms attractions in Botswana (Government Gazette, 2006)

Major attractions	Values and uses	District (location)
Gorges	Sacred; archaeological sites and some with caves or water pools	Southern (Mogonye); central (Moremi Gorge, Shoshong)
Rock art/ paintings/prints	Cultural; historical	Several places in central, Chobe, Ngamiland, Ghanzi, Kgatleng, north-east and southern districts
Hills	Sacred; cultural symbols; rock paintings/prints; relics; stonewalls	Central (e.g. Tswapong Hills, Mmashoro), southern (Otse hills), Ghanzi (Kangumere), Ngamiland (Aha, Koanaka, Tsodilo)
Caves	Cave formations; stalagmites; sinkholes	Southern (Lobatse), Ngamiland (e.g. Gcwihaba)
Hot spring	Medicinal water	Chobe – Kasane, the only known hot spring in the country
Water spring	Water flows from spring all year round	Central – Makgadikgadi pan area

---

Box 8.1   Recreational Quad-biking on the Coastal Dunes between Swakopmund and Walvis Bay, Namibia

KENNETH MATENGU

The coastline of Namibia extends along the Atlantic Ocean from north to south for about 1600 km. The vast land is a fragile ecosystem with extreme environmental variations of stark beautiful sand dunes of the Namib Desert. The Namib Desert itself defines a 80–120 km wide coastal plain between the Atlantic Ocean in the west and the Great Escarpment standing out 1000–2000 m above the sea level in the north-east of the coast (Erkkilä, 2001: 16). The unique landscape of magnificent sand dunes between Swakopmund and Walvis Bay has come to be known fondly as Namibia's recreation and leisure area, offering prime building sites for holiday homes, attracting anglers and beach lovers. Quad biking has become the famously liked, yet controversial recreational activity on the slopes and in the valleys of the dunes between the Swakop River and Lang Strand. The cold Benguela Current running parallel to the sand dunes and fog carried ashore keeps the quad bikers' eldorado cool and the sand dumb, making quad biking a comfortable fun giving off little dust from tyre frictions.

Newspapers report (see Robberts, 2006) and comments on the website (see Ecotracts, 2007) suggest that the majority of the quad bikers who damage the coastal dunes are South African and Namibian holiday makers. Several other accounts (Sheltinga & Heydon, 2005) state that in Namibia, and worldwide respectively, quad bikers lead the crowd that destroys the precious habitat irreversibly. Quad bikes are machines with the ability to negotiate all terrain enabling uncaring bikers to drive around indiscriminately, causing noise pollution, disturbing anglers and all people who want to enjoy the lonely beaches for leisure.

Quad biking classifies a niche within the Namibian tourism and leisure industry that has grown strong over the years on and off the dunes under observation. The growth and, particularly, the disastrous impact on landscapes and ecosystems have raised public discussion and debate for years. The damage caused to the environment by quad biking and 4×4 off-road vehicles critically affect the endangered endemic Damara tern birds, small reptiles and insects. In a move to better control and manage this ecologically sensitive coastal dune wall, the Ministry of Environment and Tourism erected low fence poles, hoping to keep quad bikers and off-road drivers out of the pristine environment, only to learn that off-road vandals laid the fence flat in 2004. At that time, *The Namibian Newspaper* quoted conservation officials at the Swakopmund Regional Office as saying frequent disturbances of nests by vacationing anglers and quad bikers has resulted in only half of all Damara tern eggs hatching (Barnard, 2004). Holiday makers' activities have to a large extent proven the damaging of the endemic bird species and small reptiles in that penetrated area. In December 2007, the behaviour of un-mindful visitors continued to cause distress within the conservation community again.

The exploitation of the coastal area for tourism, recreation and leisure is no doubt offering some form of employment to the local community. As such, it contributes to the economic growth of the two towns, namely the urban accommodation centre Swakopmund and the port city of Walvis Bay. Nationally branded for tourist markets overseas, this coastal sea of Namib dunes and sand also attracts travel photographers fortunately, firmly subscribing to the paradigm of non-consumptive tourism. 'Pictures' as Suvantola (2002: 183) rightly says, sustain and direct the personal memories of those involved in recreation activities, which would otherwise easily fade. In addition, photographs are indeed significant accounts to 'Others' about the 'conquered land' through recreation. It is evident that this form of recreation demonstrates elements of power that calls for more stringent control in order to build, if not enforce a sense of responsibility and ethical behaviour among both, quad-bike business operators and the visitors themselves, appreciating natural heritage in future.

Responsible tourism, although difficult to measure and define, is one that is based on mutual respect for each other and for the environment. Environmental education through initiatives such as the Road Show Initiative undertaken by the Namibian Coast Conservation and Management Project (2007a, 2007b) should be encouraged in order to provide information in a simple and yet comprehensive way to enhance people's knowledge about the cross-cutting nature of the environment. This effort might ultimately influence, if not change the behaviour among local residents, national and international tourists, alike. The conclusion to be drawn here is that if the coastal area of Namibia is seriously set to gain the degree of repute commensurate with responsible visitor conduct and respect towards environment, a transformation not only from a law enforcement point of view but also from within the tourism industry itself is required.

Community-based tourism is a mode of tourism that is initiated, managed and owned by local communities and these include landforms such as Matsieng ancient footprints (near Mochudi), Tsodilo Hill in Ngamiland, Domboshaba in north-east and Moremi Gorge in Tswapong area. Community-based tourism drive to empower local communities realize and appreciate the environment as the local economy for the local community as they irk a living and all other social benefits from it. Thus, the prime purpose is to set up sustainable societies. This means biodiversity and other natural goods being conserved for ethical and some utilitarian purposes, with resistance to demands for alternative land uses by the minority who want to serve their private business reasons (Rodriguez *et al.*, 2007). Saarinen (2006) highlights the various challenges under community-based tradition, for example the power issues – who may define the acceptable level of use (including ecological limits), which resources get sustained and by whom for whom, or what the meanings and perceptions of sustainable tourism are there for and by local cultures. Tourism in itself may entail resource, activity and community-based limits in tourism, which can be discussed in terms of the current debates of sustainable development, or environmental sustainability as coined in the UN Millennium Development Goals and in the Botswana Vision 2016, of building a nation that is prosperous and productive; that is just and secure. Conflicts have arisen between tourism resources and other sectors (e.g. agriculture) in Botswana and as tourism grows the challenges are likely to persist. The conflicts lead to stakeholders seeing others as their rivals, so each stakeholder acts to maximize single function use (de Groot, 2006). Full consultation of various stakeholders, informed decision-making on the value of ecosystems and landscapes, including the costs of their loss is essential in coming up with sustainable solutions.

The various policies such as TGLP, wildlife management, water conservation and others have a bearing on the tourism policy. In essence, the access, utilization and benefits from the land use and natural resources do affect tourism, and are affected by tourism. Thus landscapes in Botswana may be heritage, cultural, religious, wildlife habitats sites as well as expansive regions like the Kalahari landscape, whose main value is wilderness.

Conclusions

Botswana has an abundance of natural riches (landforms and wildlife) for which there are many interests. The global move is to promote geo-conservation (e.g. geoparks) – where preservation of unique geological features and heritage is promoted. The geotourism, whether geological/geomorphological or geographical, seem to be diffuse in Botswana. Geologic/geomorphic features do not receive the highest visitation numbers, rather it is wildlife (BNES, 2002). However, some compelling brochures emphasizing the uniqueness of Botswana use landscapes, for example Okavango, the jewel of the Kalahari. In this example, it may also be argued that the geographical side of ecotourism prevails (the other geo-tourism) – the environment, culture, aesthetics and well-being would be covered. Thus, the global dilemma of tourism terminology also prevails for Botswana. Some of the landscapes and landforms in the country have been recognized as cultural and natural heritage sites, but many remain unclassified on a district by district basis. The government recognizes the importance of sustaining the landscapes, ecological base and quality natural resources that provide for a wide range of uses. Decisions about natural resource allocation must be developed through a rational and open land-use planning process that integrates and recognizes a multi-functional analysis of landscape/ecological values which bring together several disciplines so as to achieve the sustainability and enhance conservation of the natural capital, as well as empower local communities.

In the current global debates, the sustainability of natural systems (bio-physical) is not only limited to human or developmental challenges. Climate change has come up as one of the main challenges. According to Scott *et al.* (2007), it impacts the resources that define the nature and the quality of the environment, thereby reducing the appeal of a wider region and tourism products. In Botswana, the perceived threats arising out of climate change (e.g. increased veldt fires, impacts on water resources, likely shifts in foraging and migration of the various wildlife, and impacts on heritage sites and visitations to parks) are areas that require more detailed research.

In dealing with landscapes and local resources, it is important to empower local communities. In the modern era, landscapes are largely seen as commodities, and thus become consumption items, dictated by economic modes. Very often the benefits of tourism resources do not trickle down to the local communities (see Mbaiwa & Darkoh and Moswete *et al.*, Chapter 11 in this book). In addition, the meaning and purpose of nature conservation must not emphasize ecological intrinsic value over cultural belonging (Bianchi, 2002). The various models of protection/conservation for heritage sites versus the consumption mode of promoting these sites as tourist attractions continue to invite challenges.

References

Albertson, A. (1998) *Northern Botswana Veterinary Fences: Critical Ecological Impacts Policies*. Maun: Okavango Wildlife Trust.

Alpizar, F. (2006) The pricing of protected areas in nature-based tourism: A local perspective. *Ecological Economics* 56 (2), 294–307.

Arntzen, J.W. (2003) *An Economic View on Wildlife Management Areas in Botswana* (Occasional Paper 10). CBNRM Support Programme, IUCN, Gaborone, Botswana.

Barnard, M. (2004) No Christmas for endangered terns. *The Namibian Newspaper*, 22/12/2004. On WWW at http://www.namibian.com.na/2004/December/national/04854E47E6.html. Accessed 11.12.2007

Bianchi, R.V. (2002) The contested landscapes of World Heritage on a tourist island: The case of Garajonay National Park, La Gomera. *International Journal of Heritage Studies* 8 (2), 81–97.

Botswana Government Gazette (2006) XLIV, 65, Gaborone, 1 September.

Botswana National Ecotourism Strategy (BNES) (2002) Gaborone, Botswana: Government Printer.

Botswana Tourism Board Act (2003) Act No. 14 of 2004. Gaborone, Botswana: Government Printer.

Buckley, R. (2006) Geotourism –publications in review. *Annals of Tourism Research* 33 (2), 583–585.

Chanda, R., Ayoade, J. and Gwebu, T. (2001) Geography of Botswana. In Department of Surveys and Mapping (ed.) *Botswana National Atlas* (pp. 1–12). Gaborone: Department of Surveys and Mapping.

Cooke, H.J. (1985) The Kalahari today: A case of conflict over resource use. *The Geographical Journal* 151 (1), 75–85.

Crofts, R.S. (1975) The landscape component approach to landscape evaluation. *Transactions of the Institute of British Geographers* 66, 124–129.

de Groot, R. (2006) Function analysis and evaluation as a tool to assess land use conflicts in planning for sustainable, multi-functional landscapes. *Landscape and Urban Planning* 75 (1), 175–186.

Deng, J., King, B. and Bauer, T. (2002) Evaluating natural attractions for tourism. *Annals of Tourism Research* 29 (2), 422–438.

Ecotracts (2007) At http://www.ecotracs.org/NOW/QuadBikes.htm#Mani%20 issue. Accessed 11.12.2007.

Erkkilä, A. (2001) Living on the land: Change in forest cover in north-central Namibia 1943–1996. University of Joensuu, Joensuu, Silva Carelia.

Department of Wildlife and National Parks (1986) *Wildlife Conservation Policy*. Gaborone, Botswana: Government Printer.

Government of Botswana (1992) *Wildlife Conservation and National Parks Act*. Gaborone, Botswana: Government Printer.

Government of Botswana (2001) *Monuments and Relics Act*. Gaborone: Government Printer.

Maher, P.T. (2007) Book review: Geotourism. *Tourism Management* 28 (2), 350–351.

Mehmetoglu, M. (2007) Typologising nature-based tourists by activity – theoretical and practical implications. *Tourism Management* 28 (4), 651–660.

Ministry of Finance and Development Planning (2003) *National Development Plan 9: 2003/04–2008/09*. Gaborone, Botswana: Government Printer.

Ministry of Trade, Industry, Wildlife and Tourism (MTIWT) (2002) *Game Ranching Policy for Botswana*. Gaborone, Bostwana: Government Printer.

Ministry of Trade, Industry, Wildlife and Tourism (MTIWT) (2004) *Community Based Natural Resource Management Policy. CBNRM Policy* (Draft 2004) Gaborone, Botswana: Government Printer.

Mowforth, A. and Munt, I. (1998) *Tourism and Sustainability: New Tourism in the Third World*. London: Routledge.

Namibian Coast Conservation and Management Project (2007a) *Review of Existing Institutional Mandates, Policies and Laws. Relation, Coastal Management and Proposal for Change* (Final Draft Report). Prepared by Southern African Institute for Environmental Assessment, Windhoek. On WWW at http://www.nacoma. org.na/Downloading/ReviewExistingPoliciesAndMandates.pdf. Accessed 26.03.2008.

Namibian Coast Conservation and Management Project (2007b) *Strategic Environmental Assessment (SEA) for the Coastal Areas of Erongo and Kunene Regions* (Draft Report, July). Prepared by DHI Group Water and Environment. On WWW at http://www.nacoma.org.na/Downloading/NACOMA_Strategic_ Environmental_Assessment_Final.pdf. Accessed 26.03.2008.

Robberts, E. (2006) Quad-bike menace continues at coast. *The Namibian Newspaper*, 4/01/2006. On WWW at http://www.namibian.com.na/2006/January/ national/06FA193A3E.html. Accessed 11.12.2007.

Rodriguez, J.P., Taber, A.B., Daszak, P., Sukumar, R., Valladares-Padua, C., Padua, S., Aguirre, L.F., Medellin, R.A., Acosta, M., Aguirre, A.A., Bonacic, C., Bordino, P., Bruschini, J., Buchori, D., Gonzalez, S., Mathew, T., Méndez, M., Mugica, L., Pacheco, L.F., Dobson, A.P. and Pearl, M. (2007) Globalization of conservation: A view from the South. *Science* 317 (10), 755–756.

Saarinen, J. (2006) Traditions of sustainability in tourism studies. *Annals of Tourism Research* 33 (4), 1121–1140.

Scott, D., Jones, B. and Konopek, J. (2007) Implications of climate and environmental change for nature-based tourism in the Canadian Rocky Mountains: A case study of Waterton Lakes National Park. *Tourism Management* 28 (4), 570–579.

Sheltinga, D.M. and Heydon L. (eds) (2005) *Report on the Condition of Estuarine, Coastal and Marine Resources of the Burdekin Dry Tropics Region. Report to the Burdekin Dry Tropics Region*. Brisbane: EPA.

Suvantola, J. (2002) *Tourist's Experience of Place*. Adelshot: Ashgate.

Waitt, G., Lane, R. and Head, L. (2003) The boundaries of nature tourism. *Annals of Tourism Research* 30 (3), 523–545.

Chapter 9

Tourism, Nature Conservation and Environmental Legislation in Namibia

SUSANNE SCHOLZ

Introduction

Namibia's young and emerging tourism industry is primarily nature based. Visitors remember the rather pristine and breathtaking diversity of landscapes with their free roaming wildlife populations as main attractions of the tourism experience. Since Independence (1990), the previously 'whites-only' tourism industry turned into an important vehicle for growth and development. In 2008, this potential competed with the mining sector in performance. The transformation of the tourism sector required the redefinition of the pre-Independence understanding of nature conservation. 'Conservation' has received new qualifications, moving away from an exclusive concept favouring the 'protection of State land' towards the holistic governance of land use in fragile environments of 'protection areas' within the paradigms of sustainability and environmental justice.

State and government regulation through post-Independence legislation addresses gaps and shortcomings inherited from the past. Reforms in legislation aim at providing communities, whose livelihoods heavily depend on the natural environment, with new opportunities for advancement. Countrywide, managerial concepts such as 'community-based tourism' (CBT) and 'community-based natural resource management' (CBNRM) have become central instruments, guiding tourism development and sustainable management of natural resources.

This chapter examines factual situations evolving from recent tourism and environmental legislation, and implementation thereof. The author presents efforts undertaken in 'land management' and 'conservation' on state-owned protected land, commercial farmland and communal lands, embedding deliberations in the politico-economic and socio-cultural context in flux. Arguments place special emphasis on answers to the question, how government approaches the political mandate and distributive challenge underlying the objective of conserving resources for the benefit of communal areas, which harbour almost 60% of Namibia's population; especially the rural poor. With a view to the topic's complexity, the contribution intentionally offers only scenarios to identified conflict formations that call for resolution.

The Legal Framework

Tourism and conservation

As outlined in government's *Vision 2030* (2004: 151), 'almost all tourists visiting the country expect a wildlife-centred experience – either through game-viewing, bird-watching, hiking, sport fishing or trophy hunting'. Wildlife experiences, coupled with Namibia's sparsely populated, spectacular arid scenery and wide-open spaces, are undoubtedly the country's greatest attraction. In today's stressful, overcrowded, rapidly developing world, natural environments are threatened and disappearing fast. Consequently, the solitude, silence and natural beauty of Namibia's landscapes become sought after commodities that have to be recognised as valuable assets. Preserving these natural assets is fundamental to developing sustainable tourism in Namibia and thereby maintaining a comparative advantage within the global market economy (Republic of Namibia, 2004: 151).

South Africa's rule over Namibia (1919–1990) did not take into consideration the development needs of the country. Policy making was oriented towards urban settlements and commercial (white) farming areas, largely neglecting the environmental sphere (Blackie & Tarr, 1999: 6). Virtually, no environmental planning existed between the country's former economic actors or across regions (Brown, 1996: 16).

From Independence, the Republic of Namibia became one of the world's first countries to incorporate an environmental and sustainable development clause (Article 95 (l) and (c)) in the National Constitution. The country's commitment to sustainable development is therefore anchored at highest level. At national and cross-sectoral level, Namibia's five-year National Development Plans, such as Namibia's National Development Plan 2 (NDP2, 2001–2006) and currently revised NDP3 (2007–2012), address the concern and need for action focusing on sustainable development. The visioning exercise *Vision 2030* also fully embraces the idea of sustainable development and calls for every Namibian to have living standards equal to those in the developed world by 2030 (Republic of Namibia, 2004). Perceived as a long-term people's manifesto, the document involved people's input from all ranks in the formulation of the content. Although considered to be 'not a Government document', it now accompanies all institutions of public and private development planning in their monitoring of performance and designing of future development programmes and projects in Namibia (Becker, 2006: 18; see Becker, Chapter 6 in this book).

Vision 2030 (2004: 150–151) acknowledges that Namibia's tourism industry is capable of contributing to wildlife conservation and biodiversity protection, and can support poverty alleviation, particularly in rural areas. The draft National Tourism Policy (2005) emphasises the necessity to develop an 'acceptable tourism' that is 'economically, socially and environmentally sustainable – tourism that contributes positively to the local and national economy, the local environment and the empowerment of local people, as well as other previously disadvantaged people. This means encouraging the sort of tourism that has the greatest chance of providing a long-term future for local communities, where ongoing market demand is most likely to sustain tourism businesses and where tourism can assist in environmental conservation' (Republic of Namibia, 2005: 5).

In order to meet outlined goals and visions, post-Independence legislation addresses shortcomings and discrepancies of the past. The former Directorate of Nature Conservation and Tourism was elevated to the Ministry of Environment and Tourism (MET), initially named Ministry of Wildlife, Conservation and Tourism from 1990 to 1994 (Asheeke & Katjiuongua, 2008: 28). New legislation aims at promoting nation-wide environmental conservation and widening the base of who benefits from tourism development (Table 9.1).

Table 9.1 Strategic framework (1990–2007): Tourism and nature conservation in Namibia (selected documents)

Year	Document	Objective	Institution
1990	The Constitution of the Republic of Namibia, Article 95(l) and (c)	To commit Government to sustainable utilisation of natural resources for benefit of all	Government (GRN)
1993	Namibia's Green Plan Environment and Development	National vision for sustainable development tabled at UNCED 1992, Rio de Janeiro To analyse environmental issues and identify actions required	Ministry of Wildlife, Conservation and Tourism (ed.: C.J. Brown)
1993	Namibia's 12 Point Plan for Integrated and Sustainable Environmental Management	Short, strategic implementation document. To set out priority actions until 1995	Ministry of Wildlife, Conservation and Tourism (ed.: C.J. Brown)

(Continued)

Table 9.1 *Continued*

Year	Document	Objective	Institution
1994	White Paper on Tourism	To guide tourism development	MET
1995	Namibia's Environmental Assessment Policy for Sustainable Development and Environmental Conservation	To guide Environmental Assessments	MET
1995	Wildlife Management, Utilisation and Tourism in Communal Areas	To provide for rural communities on communal land to manage and benefit from wildlife and other renewable resources through the formation of 'conservancies'	MET
1995	Promotion of CBT	To promote community-run tourism activities on communal land	MET
1996	Nature Conservation Amendment Act	Puts into effect the above Policy on *Wildlife Management, Utilisation (...)* and grants conservancies same rights over wildlife as freehold farmers	Office of the Prime Minister
1996	Tourism Act, consolidated draft	To provide for better coordination and regulation of the tourism industry; and to give conservancies concessionary rights over tourism activities	MET
2004	Vision 2030	To provide direction to all actors in order to reach vision 'to improve the quality of life of the people of Namibia to the level of their counterparts in the developed world, by 2030'	GRN
2004	Transformation Charter for the Tourism Sector of the Republic of Namibia	To set tourism industry standards for BBEE programmes	Federation of Namibian Tourism Associations (FENATA)

(Continued)

Table 9.1 *Continued*

Year	Document	Objective	Institution
2005	National Tourism Policy, first draft	To guide tourism planning and development	MET
2007	Tourism and Wildlife Concessions on State Land	To grant tourism and trophy hunting concessions on State land, while empowering communities	MET

Sources: The Republic of Namibia (1990, 1993a, 1993b, 1994, 1995a, 1995b, 1996, 2004), NACSO (2007a), FENATA (2004) and Weidlich (2007)

In an attempt to address past inequalities, Affirmative Action (AA) and Broad-based Black Economic Empowerment (BBEE) objectives were instrumented in the Namibian tourism industry. Another fundamental principle is the idea of equal benefits from free-roaming wildlife populations in different land tenure systems.

BBEE and the involvement of communities in tourism

Until the 1990s, Namibia's tourism industry predominantly attempted to meet the demands of the then marginal number of tourists, mostly classified as 'Visiting-Friends and Relatives (VFR)' and 'Business'. Shortly after Independence, tourism was identified as a priority sector for development and advancement. The local tourism industry, perceived to be predominantly in 'white hands' (Ashley & Barnes, 1996: 13), had to be addressed in order to widen the base of beneficiaries. The Policy on Promotion of Community-based Tourism Development (1995) promotes social and economic development in communal areas through tourism and 'recognises that due to historical inequities in the tourism industry, residents of communal areas have rarely been involved in the planning of tourism activities' (Republic of Namibia, 1995b: 3), while the formation of communal area conservancies (see below) aims at rectifying past inequalities. Furthermore, the Transformation Charter for the Tourism Sector (2004), signed by industry representatives from the public and private sectors, aims at channelling the benefits of tourism to previously disadvantaged Namibians and defines industry standards for BBEE programmes (Federation of Namibian Tourism Associations, 2004).

Benefits from free-roaming wildlife populations

During the era of South African rule, focus of governmental development efforts was primarily on freehold commercial land. In 1967, the *Nature Conservation Ordinance* (amendment 1975) was passed granting conditional ownership rights of previously state-owned wildlife to freehold farmers. Wildlife, once decimated as food competitor to livestock, gained economic value through legalised consumptive and non-consumptive uses. Together with a well-adapted rangeland management and a holistic approach to wildlife and natural resource management, this law had wide ranging effects on wildlife populations which drastically increased in the following decades. A large proportion of Namibia's wildlife developed outside protected areas and formed the basis for a growing wildlife-based industry on freehold land.

On the contrary, communal area residents were legally deprived from exploiting wildlife resources on their land, until legislative adjustments were introduced in the mid-1990s. Wildlife conservation was conducted on the principle of protecting resources from people and prosecuting poaching. With tourism, rural area residents had to tolerate the intrusion of (adventure) tourists, but received little individual or collective benefits. The independent Namibian Government sought to rectify these past inequalities and inconsistencies in its approach to commercial and communal farmland and their tenants. Based on earlier success in privatising wildlife on commercial land, wide reaching policies and legislation were introduced in the mid-1990s, paving the way for a new era of nature conservation and tourism development on communal land. The *Nature Conservation Amendment Act* of 1996 grants communal area residents equal rights, obligations and responsibilities in wildlife and natural resource utilisation.

The creation of economic incentives is perceived as a key to maintaining wildlife populations and habitats on communal lands. Namibia has been a protagonist in the SADC region in its endeavours to devolve natural resource rights to the community level, while favouring CBT development.

Tourism and Nature Conservation in Protected Areas

With 16.5% (2008) of Namibia's surface area comprising national parks and game reserves, the country exhibits a strong commitment to conservation for tourism. Main nature-based tourism destinations include the Namib-Naukluft National Park with its diverse desert landscapes and the claimed 'highest dunes in the world' around Sossusvlei, not to exclude the Etosha National Park with its abundance of wildlife around the usually dry Etosha Pan. These two national parks were proclaimed as early as 1907 by German colonial power. The Republic of Namibia confirmed and reinforced inherited protected areas and expanded the network with additional proclamations and innovative forms of conservation (see Becker, Chapter 6 in this book). Today, Namibia's state-controlled Protected Area Network (PAN) comprises 21 parks and reserves. The 'visitor management approach' to the country's national parks may be perceived as stringent and conservative in respect of strict opening hours and the restriction of game drives to designated roads.

In pre-Independence Namibia, nature conservation was largely seen as an issue of 'parks and wildlife' (Brown, 1996: 17) and conservation was pursued on the basis of excluding people living in the vicinity of parks. In line with the global paradigm shift that conservation can only be successful 'with the people', 'parks and their neighbours' are emerging as the new guiding principle in Namibia. This approach is best illustrated with the Etosha National Park, where a stronger cooperation between the Park and adjoining freehold farms, communal land and other entities is envisaged for the future (see Sandpaper, 2007). In order to maximise tourism potential and revenues, the Policy on Tourism and Wildlife Concessions on State Land (2007) provides unexplored opportunities that regulate the granting of tourism and trophy hunting concessions on State land. The policy includes game parks, protected as well as communal areas, and strives to empower communities living in these areas through tourism. Concessions are to become part of the future *Parks and Wildlife Management Act*, which will replace and consolidate current legislation (Weidlich, 2007). Following the SADC initiative, Namibia also embraces the concept of transboundary conservation, with the first park, the Ai-Ais/Richtersfeld Transfrontier Park with South Africa, launched in 2003.

Namibia's protected areas are undoubtedly the country's most important tourism destinations. The presence of a national park adds value and favours tourism development on neighbouring lands as well (Ashley & Barnes, 1996: 11). A recent assessment by the WTTC (2006: 49) foresees considerable potential in the expansion of tourism enterprises in and around national parks as well as the proclamation and development of trans-frontier parks. Namibia's conservation efforts on state-protected land are highly recognised internationally (WTTC, 2006: 48). However, Barnard criticised in 1998 the fact that some hotspots of endemism, such as the western escarpment, have not become protected areas (Blackie & Tarr, 1999: 13).

Tourism and Nature Conservation on Freehold Commercial Farmland

Major parts of Namibia's territory are conventionally only suitable for nomadic or rotational grazing. Limiting factors include poor surface water availability, erratic rainfall and thin, infertile soils (Barnard, 1998: 35). Extensive livestock farming forms the basis for survival on freehold commercial farmland throughout Namibia, with cattle dominating the tree-and-shrub savannahs of central and northern Namibia, and sheep being best adapted to the dryer savannah zones and Nama Karoo in the south and south-west.

The South African Odendaal Commission of 1963 sought to enforce the ideologies of Apartheid on a nationwide scale. The envisaged spatial and social 'segregation of the races' was implemented with great vigour in the years 1964–1966, and led to the division of Namibia's territory into 'whites-only' commercial farms and communal 'black homelands'; with both categories covering just about 43% of the country's surface. Governmental development efforts focused primarily on the 'white heartland'. Heavily subsidised during the 1970s, the application of scientific and technological solutions was promoted. Following the promulgation of the *Nature Conservation Ordinance* in 1967/1975, granting ownership of wildlife to commercial farmers, farm operations diversified to include wildlife products. Subsequently, wildlife populations increased by approximately 70% during the period 1972–1992 (Barnes & De Jager, 1995: 9).

The Republic of Namibia chose to maintain the country's division into communal and freehold areas. The 'veterinary cordon fence' in northern Namibia, erected in the mid-1960s in order to protect commercial farms south of the fence from foot and mouth disease, was maintained in order to secure Namibia's access to international export markets. With accelerating tourism development from the 1990s, an increasing number of commercial livestock farmers shifted their main, if not only, source of income to wildlife management and tourism activities. Economic benefits include non-consumptive wildlife viewing as part of the tourism experience, paralleled by consumptive uses such as venison production, live game capture and sale, as well as recreational trophy hunting. The mushrooming of new guest farms and lodges offering photo and hunting safaris underline this trend (Ashley & Barnes, 1996: 6).

Since 1991, a number of commercial area conservancies (21 in 2007) emerged throughout the country, aiming at the managing of wildlife populations migrating over farmland in a joint and sustainable manner. Conservancy members agree to allocated wildlife quotas determined by the conservancy committee (Albl, 2001; Conservancy Association of Namibia, 2007). In the early years of the implemented concept, a further driving force was the perception, that such a cooperation could possibly offer land tenure security. Not recognised by the government, today resettlement schemes on individual farms within delimited conservancies jeopardise conservancy efforts. Privately owned nature and game reserves running predominantly low impact–high return tourism, emerged as a further business approach to conservation and tourism on commercial farmland (e.g. NamibRand Nature Reserve, currently 172,200 ha; Gondwana Cañon Park, 112,000 ha).

The past decades witnessed the diversification of rural economies on freehold farmland, the owners of which added consumptive and non-consumptive wildlife tourism to their income generating activities. Today, wildlife and livestock management combined with tourism ventures proves to offer viable revenue options on freehold land.

Tourism and Nature Conservation on Communal Land

One of the key challenges to nature conservation and tourism policy development in Namibia has been posed by the need to distribute potential benefits of natural resources and tourism more evenly across the country and its people.

Communal areas reflect ethnic 'homelands' of the past and are primarily found in the northern border regions of the country. Today, some of these areas, such as Namibia's north-west, are part of the country's most attractive tourism destinations. Valuable wildlife resources occur in these less densely settled areas. Species of importance to conservation in the semi-arid and sub-humid north-east include elephant, buffalo, hippopotamus, sable, roan, lechwe, sitatunga, lion, leopard and wild dog. The arid north-west is populated by desert-adapted elephant, black rhino, mountain zebra, as well as the more common springbock, kudu and oryx, living in attractive landscapes throughout (Barnes *et al.*, 2001: 3).

Quite a number of people and governing bodies exert control over these peripheral areas, ranging from individual families, traditional authorities, regional and central government institutions. Customary law is increasingly questioned by local people receiving greater access to civil law. These communal areas are affected by unemployment, poverty, poor standards of health and education resulting in slow and uneven economic growth. Agro-pastoralism forms the basis for survival in the highly populated north and north-east of the country and rural dwellers remain heavily dependent on the primary production and harvest of natural resources. Unsustainable land-use practices have led to overgrazing and soil erosion. They include illegal fencing, inconsiderate land clearance, uncontrolled bush fire and the displacement and killing of wildlife. The pressure on the drought-prone natural environment is further accelerated by fast growing populations (Brown, 1996: 18).

With outlined legislative amendments in the mid-1990s, communal area residents have acquired partial rights to common property resources such as wildlife. Through the formation of spatially delimited 'communal area conservancies', members become entitled to manage, use and benefit from resources and tourism potential on 'their' land. The inherent Namibian approach makes provision for conservancies to emerge through people's own initiative, and communities can define themselves and their boundary. In order to qualify for registration with the MET, conservancies require defined boundaries of jurisdiction, a defined membership, a representative management committee, a legally recognised constitution that makes provision for the development of a wildlife management strategy and an equitable benefit distribution plan (NACSO, 2007a). A key assumption underlying the rationale of the conservancy concept is that wildlife use and tourism offer comparative advantages over traditional forms of land use. Conservancies constitute economic and social management units that may considerably diversify rural communities' livelihoods while contributing to conservation. The first four communal area conservancies were gazetted in 1998, increasing to 50 by the end of 2006, with a further 20 currently being formed (NACSO, 2007b; Figure 9.1).

The past decade has been characterised by a new openness to conservation and concurrent use, taking the form of CBNRM and CBT in communal areas of Namibia (see Saarinen, Chapters 5 and 15 in this book). Both concepts emerged before Independence and were institutionalised in the 1990s.

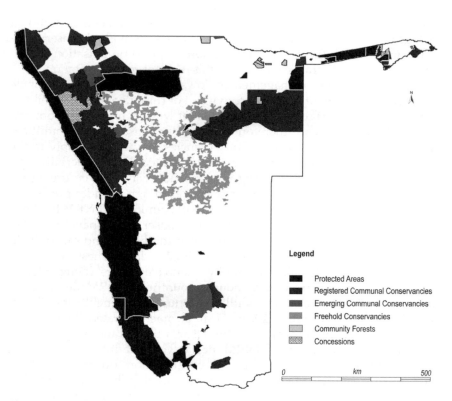

Figure 9.1   Conservancies and protected areas

Community-based Natural Resource Management

Namibia's CBNRM has its early roots in 1983, when the local NGO 'Integrated Rural Development and Nature Conservation (IRDNC)' developed a community game guard system in the arid north-west of the country, in order to protect endangered species such as desert elephant and black rhino from extinction (IRDNC, 2007). CBNRM programmes aim at guiding communities to manage and use natural resources in sustainable and productive ways. Essential cornerstones of the concept include the reliance on community consensus, the use of incentives and the decentralised administration of resources (Coombes, 2007: 61). The holistic and integrated approach strives towards natural resource management, enterprise development and institutional progression (Long, 2004: xiv). Namibian CBNRM is a national programme and represents a joint venture between government and non-governmental institutions, communities, community-based organisations and development partners (NACSO, 2007a). The Namibian Association of Community-based Support Organisations (NACSO) formed in the late 1990s has helped to integrate and improve coherency within the various local CBNRM programmes.

Government utilises conservancies as a means to allocate conditional resource and tourism rights to rural communities. Individual conservancy communities, with the assistance of various support organisations, identify and define best practice resource utilisation for implementation.

Successful wildlife management follows the premise of 'sustainable off-takes'. Government is not yet convinced of rural dwellers' capabilities pertaining to the sustainable management of wildlife resources. Consumptive wildlife quotas are therefore allocated to conservancy committees based on regular game counts. Conservancies may further apply to the MET for trophy hunting status (including protected and specially protected wildlife), whereby hunting is performed via joint venture agreements with private sector operators selected through a tender process by the conservancy. Although the initial focus of Namibian CBNRM has been on wildlife (Barnes *et al.*, 2001: 3), a number of conservancies have expanded programmes beyond wildlife in order to control the use and sale of 'veld products' such as the medicinal plant 'devil's claw'. These duties are usually delegated to female 'community resource monitors' (Long & Jones, 2004: 45). As opposed to some neighbouring countries' CBNRM programmes, for example Zimbabwe, where wildlife revenues also benefit rural district councils (Jones & Murphree, 2004: 87), all income generated from wildlife use remains with the Namibian communal area conservancies.

Namibia's broad-based CBNRM has developed into more than a national resource management and conservation programme. It is increasingly regarded as a rural development strategy, helping to diversify rural economies and people's livelihoods. Innovative land-use schemes supplement traditional farming in order to generate income, new jobs and to provide skills and expertise.

The concept further strives towards rural empowerment through 'increasing capacity in decision-making, problem solving and conflict resolution; strengthening links between communities, traditional authorities, local and national government and providing information and skills that can create income and develop local infrastructure' (NACSO, 2007a).

CBNRM is mostly driven by communal area conservancies. However, this is only one example how rural development based on sustainable resource management can occur. In 2001, the Namibian Government extended its CBNRM programme in order to have community forests registered under the *Forest Act*. The first 13 community forests were gazetted in early 2006 (NACSO, 2007b: 14; see Figure 9.1).

Community-based tourism

The first Namibian Community Based Tourism Enterprises (CBTE), such as the Spitzkoppe campsites or the Aba-Huab camp and guided walking trails at the rock engraving sites of Twyfelfontein in north-western Namibia, emerged before Independence, and before the 'communal area conservancy' concept was institutionalised. Twyfelfontein has been proclaimed the first Namibian World Heritage site in 2006 and forms part of the country's major tourist attractions. The majority of CBTEs are, however, located in remote areas such as Namibia's extreme north-west (Kunene Region) or the Caprivi in the east, off the 'standard tourism circuit' operating on revenues from individual travellers. Recent trends suggest an increased interest by overseas tourists to experience (authentic) cultural tourism in Namibia (Asheeke & Katjiuongua 2008: 63).

According to the World Travel and Tourism Council (WTTC, 2006: 49), Namibia has been a pioneer in the development and implementation of community-based models for tourism management. CBT products include campsites, donor-funded community lodges, cultural villages, tour guide services and craft outlets. The Namibia Community Based Tourism Assistance Trust (NACOBTA) serves as the lead agency for all CBTE (Box 9.1). The non-profit membership organisation supports communities in their efforts to develop and operate tourism enterprises profitably and in a sustainable manner.

Box 9.1:   The Namibia Community-based Tourism Assistance Trust

JARKKO SAARINEN

The NACOBTA was initiated in 1995 by local communities. It is supported by the Namibian government and as a non-profit organisation it aims to improve the living standards of people by developing sustainable CBT and especially joint ventures integrating private sector investors with local communities (Massyn, 2007). Based on its vision, NACOBTA's goal is to improve the living standards of the people and contribute to poverty reduction, the reduction of the income disparity and overall social stability in rural (communal) areas of Namibia (see NACOBTA, 2008).

The specific objectives of the NACOBTA are to

- ensure the success of CBTE and sound collaboration with the mainstream tourism industry;
- source and disseminate relevant information to CBTEs;
- facilitate needs-based, purpose-driven services;
- enhance the capacity of service providers; and
- advocate at policy level in terms of the CBT sector.

To achieve these objectives, the NACOBTA provides supporting services, such as staff training, business plan advice, marketing, funding and lobbying to its members. The members can range from campsites, tour operators, guides and community lodges to cultural villages, craft outlets, information offices and conservancies with joint ventures (NACOBTA, 2008).

The often outstanding beauty of relatively undisturbed landscapes offers communities competitive advantages in attracting international visitors seeking solitude and a nature-centred experience. Recovering wildlife populations on communal lands add to the experience and include high-value species such as desert elephant, black rhino and hippopotamus. In attractive areas, tourism enterprises often take the form of partnership agreements with the private sector through concessionary rights.

With limited opportunities to access the cash economy and inherent difficulties faced by agriculture in a semi-arid environment, CBNRM and CBT offer significant opportunities towards poverty alleviation and diversification of livelihoods in Namibia's communal areas while contributing to conservation. In 2006, communal area conservancies covered 14.4% of Namibia's land surface. Together with national parks and game reserves (16.5%), commercial area conservancies (6%) and a further 1.3% under concessions and community forests (NACSO, 2007b: 11; see Figure 9.1), an impressive land surface of approximately 38% is under some form of conservation management. New land-use categories created in the last decade have significantly altered the country's economic and conservation landscape. Tourism is regarded as the largest income generator for communal area conservancies and their creation benefits biodiversity conservation and rural development through CBNRM and CBT.

Community Involvement in Conservation and Tourism: A Success Story?

Before the implementation of the conservancy concept and its relating CBNRM and CBTE initiatives, rural dwellers had little or no incentive to tolerate wildlife on their land. Theoretically, the concept creates an opportunity to compensate for wildlife damages incurred on communal land. Enhanced levels of awareness and positive attitude towards conservation and tourism can be expected as results. Indeed, general perception is that poaching incidences have declined, while wildlife populations have increased (Long, 2004; NACSO, 2007b: 16f). In an effort to reintroduce wildlife species after years of extinction and as a gesture of support and trust, private and public sector initiatives have further donated wildlife to selected conservancies. Barnes *et al.* (2001: 18) assessed five well-established conservancies and concluded that the existence of free-roaming wildlife populations on conservancy land is a very significant factor towards economic efficiency and financial viability of conservancies.

From a socio-economic perspective, conservancies have offered formerly unknown development opportunities (and hopes) to rural residents. A number of conservancies have become financially self-sufficient. Positive examples primarily include conservancies that have entered into private sector joint venture lodge or hunting agreements. Lodges also pay user fees for scenic drives on conservancy land. Joint venture agreements foresee that local people benefit through concession fees, wages, formal training and other benefits such as increased access to information. Collective conservancy income from CBT and CBNRM is invested into community projects for the benefit of all, or alternatively paid out to households. Individuals who manage to access the anticipated benefit streams are well positioned to reach financial stability, personal fulfilment and empowerment.

Successes of community-based enterprises are challenged by a number of environmental and socio-economic dimensions. Growing wildlife populations – a sign of successful conservation efforts – have led to increased incidents with problem animals on communal land. These include livestock losses, damage to boreholes and water installations, crop damage, attacks and death of humans. Donor-funded human–wildlife conflict compensation schemes are currently being piloted in the Kunene and Caprivi Regions in order to maintain a positive attitude towards wildlife and tourism (Long, 2004: 115). Innovative ways are needed to address such conflicts, principally by creating and facilitating opportunities for generating economic value from wildlife rather than by payment of compensation (Republic of Namibia, 2004: 151).

'Decentralisation' is a key to successful CBNRM (Coombes, 2007: 61, see also Becker, Chapter 6 in this book). However, conservancies do not yet have full rights to determine off-take levels of wildlife. Quota allocations for consumptive wildlife exploitation are set by the MET, and it appears that no coherent policy is followed. The MET is addressing the issue and is increasingly attempting to devolve more decision-making to the conservancy level (Long & Jones, 2004: 35).

A number of conservancies have undertaken land-use planning and zoning for various activities. To date, conservancies do not have any legal basis to enforce zoning decisions, particularly when outsiders intrude the conservancy area for own use. With further enforcement of the *Communal Land Reform Act* (2002), empowering Communal Land Boards to take decisions on the allocation and designation of land, this shortcoming may be addressed, provided that conservancies gain strong recognition and work closely with Traditional Authorities (Long & Jones, 2004: 35). Long-term success of CBTE and CBNRM is dependent on clearly defined ownership rights. The lack of both, group land tenure security and private ownership of natural resources for communal area residents, counter development. Access to bank loans for small- and medium-sized enterprises (SMEs) are impeded by missing bank guarantees. Private sector investment is hampered by lack of investor confidence in ownership or user rights on communal land. These issues require urgent attention and need to be resolved.

Prior to and in the years following Independence, communal area residents were affected by lack of access to information, education, participation and training for decision-making in environmental management (Brown, 1996: 17) and tourism development, thereby failing to create an environment for democratic governance. Achieving sustainable tourism and natural resource management is ultimately dependent on effective decision-making at local level by resource users themselves. In Namibia, a broad spectrum of capacity building and business skills development is required, if conservancies and their activities are to function effectively. Any mechanism that removes the need for bureaucratic decision-making, and which places responsibilities and obligations at resource user level, is to be welcomed (Long & Jones, 2004: 34f).

Day-to-day operational challenges include low levels of literacy, democratic deficits and poor participation in decision-making, low levels of understanding of business ethics, elite capture of benefits or simply the challenge of finding common grounds in culturally divided societies. Institutional challenges include overstretched capacities by technical and financial support organisations, as well as more favourable market conditions for livestock versus wildlife management.

Namibian conservancies and their CBT and CBNRM activities are designed as a rural development strategy. However, the majority of conservancies remain dependent on external financial and technical help. Economic sustainability reducing vulnerability and increasing widespread resilience of rural dwellers will always represent a major challenge. For long-term success, it is imperative that CBTE and CBNRM are financially viable and are based on open market business principles.

Vorlaufer (2007: 29) concludes in his research undertaken in various conservancies in Namibia, that 'it is doubtful that tourism [and CBNRM, *the author*] will generate enough income to satisfy the expectations of the people and to compensate for the costs of biodiversity conservation'. He further argues, that biodiversity and wildlife conservation in particular can be successful only, if they will be continuously subsidised by foreign donors and the international community. Despite their well-documented benefits, Namibia's communal area conservancies remain a work in progress (Long, 2004).

Nature and Wildlife Tourism in Namibia: Geared for Growth?

The future of tourism development appears promising for Namibia. In 2006, the World Travel and Tourism Council identified Namibia as the 13th fastest growing tourism destination in the world, with the potential to become one of Africa's leading travel and tourism economies. Further, the WTTC (2006) attested Namibia to be a leader, both in Africa and the world, in its approach to conservation and community involvement in natural resource management. According to the United Nations, Namibia is Africa's top performer in Global Environment Facility (GEF) programmes (WTTC, 2006). Namibian tourism practices are generally perceived as sustainable and well established in state-run national parks and privately owned nature reserves, guest farms and lodges. Going forward, the greatest challenge to successfully implement nature conservation and tourism objectives lies within the development of communal areas and its settlements, accommodating the majority of Namibia's population.

Since Independence, government's development efforts focus on these areas affected by poverty and degradation. The incentive-based conservancy concept is a mechanism to achieve governmental goals of sustainable development. Conservancies with their CBNRM and CBT activities are firmly entrenched in national development plans and legislation, since they are regarded as critical success factors to poverty alleviation in communal areas of the developing world. Natural assets are designated to act as 'sustainable sources of wealth creation, vehicle of political empowerment, and avenues of integration into the national and global economies'

(World Resources Institute, 2005). Economic and social sustainability in communal areas will continue to be based on environmental resources and 'environmental income' (World Resources Institute, 2005) will always remain an important source of livelihood for the rural poor. The overall and most significant question remains, whether Namibia's community-based programmes will successfully contribute to broader national development objectives such as poverty reduction and economic growth (Long, 2004: xv). The overriding principle depicts that sustainable behaviour by individuals and communities can only be expected and achieved, if acceptable living standards have been reached.

The global paradigm of sustainable tourism is based on the principle that tourism development should be environmentally, economically and socially sustainable. In Namibia, the main legislative post-Independence focus of 'acceptable', meaning sustainable tourism development, is on the socio-economic sphere, that is, poverty reduction and raised living standards for the majority of the Namibian citizens. From this perspective, a truly sustainable tourism development has to make provision and include previously disadvantaged rural communities into the mainstream Namibian tourism industry in order to ensure that this potentially vulnerable economic sector continues to prosper in future.

References

Albl, S. (2001) Conservancies auf kommerziellem Farmland in Namibia – ein Beitrag zum Erhalt der Biodiversitaet in Savannenzonen? Unpublished MA thesis, Diplomarbeit, Universitaet Trier. Fachbereich VI: Geographie/ Geowissenschaften.

Asheeke, J.W. and Katjiuongua, O. (2008) *Development of Sustainable Tourism* (Country Report, Namibia). Sustainable Tourism Workshop, Birchwood Hotel, Johannesburg, South Africa, 30 November to 1 December 2007. Windhoek: FENATA/NACOBTA.

Ashley, C. and Barnes, J. (1996) *Wildlife Use for Economic Gain. The Potential for Wildlife to Contribute to Development in Namibia.* (DEA Research Discussion Paper 12). Windhoek.

Barnard, P. (ed.) (1998) *Biological Diversity in Namibia: A Country Study.* Windhoek: Namibian National Biodiversity Taskforce.

Barnes, J.I. and De Jager, J.L.V. (1995) *Economic and Financial Incentives for Wildlife Use on Private Land in Namibia and the Implications for Policy.* DEA Research Discussion Paper 8). Windhoek.

Barnes, J., MacGregor, J. and Weaver, C. (2001) *Economic Analysis of Community Wildlife Use Initiatives in Namibia* (DEA Research Discussion Paper 42). Windhoek.

Becker, F. (2006) Development of space in independent Namibia: Deliberations on the projection of politico-economic and socio-cultural transformation. In H. Leser (ed.) *The Changing Culture and Nature of Namibia: Case Studies* (pp. 17–35). The Sixth Namibia Workshop, Basel 2005. In honour of Dr Hc. Carl Schlettwein (1925–2005). Basel: Baseler Africa Bibliographien.

Blackie, R. and Tarr, P. (1999) *Government Policies on sustainable development in Namibia* (DEA Research Discussion Paper 28). Windhoek.

Brown, C. (1996) The outlook for the future. In P. Tarr (ed.) *Namibia Environment I* (pp. 15–20). Windhoek: Ministry of Environment and Tourism.

Conservancy Association of Namibia (CANAM) (2007). At http://www.canam.iway.na. Accessed 12.2007.

Coombes, B. (2007) Defending community? Indigeneity, self-determination and institutional ambivalence in the restoration of Lake Whakaki. *Geoforum* 38, 60–72.

Federation of Namibian Tourism Associations (FENATA) (2004) *Transformation Charter for the Tourism Sector of the Republic of Namibia.* Windhoek: FENATA.

Integrated Rural Development and Nature Conservation (IRDNC) (2007) At http://www.irdnc.org.na. Accessed 12.2007.

Jones, B.T.B. and Murphree, M.W. (2004) Community-based natural resource management as a conservation mechanism: Lessons and directions. In B. Child (ed.) *Parks in Transition* (pp. 63–103). London: Earthscan.

Long, S.A. (ed.) (2004) *Livelihoods and CBNRM in Namibia: The Findings on the WILD Project.* Final Technical Report of the Wildlife Integration for Livelihood Diversification Project (WILD), prepared for the Directorates of Environmental Affairs and Parks and Wildlife Management, the Ministry of Environment and Tourism, the Government of the Republic of Namibia, Windhoek.

Long, S.A. and Jones, B.T.B. (2004) Contextualising CBNRM in Namibia. In S.A. Long (ed.) *Livelihoods and CBNRM in Namibia: The Findings on the WILD Project* (pp. 25–40). Final Technical Report of the Wildlife Integration for Livelihood Diversification Project (WILD), prepared for the Directorates of Environmental Affairs and Parks and Wildlife Management, the Ministry of Environment and Tourism, the Government of the Republic of Namibia, Windhoek.

Massyn, P.J. (2007) Communal land reform and tourism investment in Namibia's communal areas: A question of unfinished business. *Development Southern Africa* 24 (3), 381–392.

NACOBTA (2008) At http://www.nacobta.com.na/index.php. Accessed 5.04.2008.

Namibia Association of CBNRM Organisations (NACSO) (2007a) At http://www.nacso.org.na. Accessed 12.2007.

Namibia Association of CBNRM Organisations (NACSO) (2007b) *Namibia's Communal Conservancies: A Review of Progress 2006.* Windhoek: NACSO.

Republic of Namibia (1990) *The Constitution of the Republic of Namibia.* Windhoek.

Republic of Namibia. Ministry of Wildlife, Conservation and Tourism/Brown, C.J. (ed.) (1993a) *Namibia's Green Plan Environment and Development.* Windhoek.

Republic of Namibia. Ministry of Wildlife, Conservation and Tourism/Brown, C.J. (ed.) (1993b) *Namibia's 12 Point Plan for Integrated and Sustainable Environmental Management.* Windhoek.

Republic of Namibia. Ministry of Environment and Tourism (MET) (1994) *White Paper on Tourism.* Windhoek.

Republic of Namibia. Ministry of Environment and Tourism (MET) (1995a) *Namibia's Environmental Assessment Policy for Sustainable Development and Environmental Conservation.* Windhoek.

Republic of Namibia. Ministry of Environment and Tourism (MET) (1995b) *Promotion of Community Based Tourism* (Policy Document 9). Windhoek.

Republic of Namibia. Ministry of Environment and Tourism (MET) (1996) *Tourism Act.* (Consolidated Draft). Windhoek.

Republic of Namibia. Office of the President (2004) *Namibia Vision 2030. Policy Framework for Long-term National Development.* Windhoek.

Republic of Namibia. Ministry of Environment and Tourism (MET) (2005) *A National Tourism Policy for Namibia* (First Draft). Edinburgh: International Tourism Consultancy Yellow Railroad.

Republic of Namibia. Ministry of Environment and Tourism (MET) (2007) *National Tourism Policy for Namibia* (Draft). Windhoek.

Sandpaper (2007) *The Newsletter of the Strengthening the Protected Area Network (SPAN) Project*. Windhoek: Ministry of Environment and Tourism.

Vorlaufer, K. (2007) Kommunale Conservancies in Namibia: Ansätze der Biodiversitätssicherung und Armutsbekämpfung? *Erdkunde* 61, 26–53.

Weidlich, B. (2007) New policy for tourism and wildlife concessions. *The Namibian Newspaper* (25.07.2007). Windhoek.

WTTC (2006) *Namibia. The Impact of Travel & Tourism on Jobs and the Economy.* Oxford: WTTC.

World Resources Institute (2005) On WWW at http://www.grida.no/wrr/046.htm. Accessed 12.2007.

Chapter 10

Transfrontier Conservation and Local Communities

MAANO RAMUTSINDELA

Introduction

> The objectives of this Treaty shall be to (a) foster trans-national collaboration and co-operation between the Parties which will facilitate effective ecosystem management in the area comprising the Transfrontier Park; (b) promote alliances in the management of natural and cultural resources by encouraging social, economic, responsible tourism and other partnerships between the Parties, including the private sector, local communities and non-governmental organisation; (c) enhance ecosystem integrity and natural ecological processes by harmonising environmental procedures across international boundaries and striving to remove artificial barriers impeding the natural movement of wildlife. (Treaty of the Ai-Ais Transfrontier Park, 2003: 8)

The above quotation summarises most of the public images of transfrontier conservation areas and their relationships with local communities. Those relationships are mediated through local economic development initiatives and various kinds of community-based natural resource management schemes. This broad scope is articulated by a definition of transfrontier conservation areas that not only facilitates the creation of new and expanded nature conservation areas, but also embraces the linking of nature conservation to local economic developmental needs and aspirations, particularly in remote rural areas. Transfrontier conservation areas (hereafter TFCAs) therefore attempt to demonstrate the feasibility of the notion of 'benefit beyond boundaries', which was aptly captured by the theme of the World Parks Congress held in Durban in 2003. Explicit in that notion is the view that nature conservation should benefit both the environment and the people. The notion seeks to bridge two main ideological camps. The first camp is made up of diehard conservationists who believe that nature conservation strategies should be concerned with the protection of nature. Trying to pursue multiple goals through nature conservation, the camp argues, does not augur well for the urgency required for the protection of nature (Brandon *et al.*, 1998; Terborgh, 1999). The second camp holds the view that strategies for nature conservation are most likely to succeed and gain support from a wide range of people and institutions if it addresses socio-economic development or is seen to be doing so (Adams & Hulme, 2001; McNeely & Pitt, 1985).

TFCAs articulate the spirit and purpose of the second camp; hence they are premised on the creation of conservation areas for the benefit of nature and the citizens of participating states. This, it has been argued, can be achieved by bringing together formally protected areas, freehold land, state and communal land under a common regime. It is anticipated that once such a regime is established, land use planning in and around protected areas forming TFCAs would be placed under the control of one legal authority or institution. As TFCAs are created across state borders, the amalgamation of the mosaic of land use takes place at a supranational scale. This scale is considered favourable for managing ecosystems that are critical for life on earth, and for protecting them against the whims of politics, especially in developing countries where there are considerable doubts about the viability of states. The supranational scale is also considered advantageous for international tourism with assumed local spin-offs.

The place of local communities in TFCAs is not as simple as the Treaty above describes it. The complex process of creating TFCAs, the diverse nature of TFCAs and the range of stakeholders involved in these initiatives impinge upon the place of local communities in TFCAs. Moreover, communities involved in or affected by TFCAs are different and do not have the same legal backing and assets that are crucial for their engagement with TFCAs. Under these conditions, the place of local communities in TFCAs is not pre-given but is negotiated throughout the process of creating these conservation areas and the development of appropriate management structures. In this chapter, 'the place of TFCAs' is used as a metaphor to underscore various ways in which local communities relate to cross-border conservation projects and the processes which shape the relationship between the two. The chapter analyses the place of local communities in TFCAs from three related angles. Firstly, it unpacks the conceptions of local communities as presented in official discourse so as to understand constructions of 'the community'. Secondly, it uses case studies to demonstrate the ways in which local communities are brought into, and participate in, TFCAs. Thirdly, it presents the TFCA beneficiation framework and its bearings on local communities. Some brief comments on the creation of TFCAs could aid our understanding of the theme of this chapter.

Creating TFCAs

Contemporary TFCAs owe their existence to the revival of bioregional planning as a comprehensive approach to biodiversity protection and management under stiff competition for land use, and as a response to alarming reports about the future of the earth and its inhabitants. The main focus of bioregional planning is the creation of 'a regional-landscape scale of matching social and ecological functions as a unit of governance for future sustainability that can be flexible and congruent still with various forms of governance found around the world' (Brunckhorst, 2002: 8). Those regional landscapes are important in that they form the basis on which TFCAs are created and managed. In other words, the ecological rationales for bioregions or regional landscapes and TFCAs are basically the same; they all seek to protect the integrity of ecological systems. Where such ecosystems have been disrupted or destroyed, TFCAs aim to re-establish them. This begs the question of how this could be done in reality. Methods used to create large-scale bioregions include the creation of ecological corridors, which are meant to enhance the connectivity of land-scapes. Connectivity in this context means 'the extent to which a species or population can move among landscape elements in a mosaic of habitats types' (Hilty *et al.*, 2006: 90). Interest in corridors can be linked to concerns that the degradation of the physical environment severely affects habitat connectivity, and therefore the survival of species (Bennett, 2003). Where corridors are necessary but do not exist, efforts could be made to create them. This leads to two main categories of corridors, namely, unplanned and planned corridors.

Although unplanned or de facto corridors often exist for other pur-poses, they can promote the movement and survival of species, as is the case with vegetation strips along roads, agricultural landscapes, ditches and so on (Bennett, 1990). For their part, planned corridors such as green-belts are designed to promote connectivity for biodiversity or, as the quotation above suggests, to enhance ecosystem integrity and natural eco-logical processes. Planned corridors are crucial to the creation of TFCAs and are a common feature in wildlife areas where the need for migratory routes and managing human–wildlife conflict is highest. They are by and large created by acquiring land separating habitats or by realigning land use to the goals of nature conservation.

For TFCAs, the creation of corridors is meant to facilitate the amalga-mation of contiguous protected areas, especially where these areas are separated by other land-use types. It is important to note that some such areas are communal land, which is useful for linking local communities to TFCA projects, as I will be showing below. The most common method for creating TFCAs is the joining together of adjacent protected areas, particu-larly national parks. This method has actually spurred the global search for TFCA sites. Zbicz and Green (1997) estimated that there were 136 transboundary complexes worldwide, which meant the existence of contiguous protected areas that could translate into 136 TFCAs. The implication for using protected areas as the core of TFCAs is that the number of TFCAs becomes dependent on that of protected areas. The result is the need to create more protected areas in order to increase the possibility of more TFCAs, as the case of the Great Limpopo Transfrontier Park (GLTP) below makes clear. More importantly, using protected areas as a foundation for TFCAs severely limits the ways in which local communities are accommodated into these conservation areas.

Conceptions of Communities in TFCAs

Despite its elusiveness, the concept of community still permeates various studies, including those on people and natural resources. In society–environment studies, the concept found expression in community-based natural resource management, which became prominent in the 1980s. Reasons for the adoption of the concept include attempts to articulate new forms of natural resource governance in line with the ideals of democracy; to manage local residents and to structure and monitor resource usage; to gain access to resources under different tenure systems and so forth (Adams & Hulme, 2001; Hutton *et al.*, 2005). Against this backdrop, the community in natural resource management generally denotes residents who are located within or along the borders of designated natural resource areas. It has become a synonym for 'the local' in the hierarchy of scales. The same conception of the community is found in nature conservation areas, and is expressed through notions such as community-based conservation. Community-based conservation implies the participation of local residents in the protection of nature and the use of, and benefits from, natural products (see Becker and Scholz, Chapter 9 in this book).

Conceptions of TFCAs are imbued with the 'localness' of communities. It is therefore not surprising that, in TFCA treaties, local communities are defined as groups of people living in and adjacent to the area of the transfrontier park who are bound together by social and economic relations based on shared interest. It is clear from this definition that local communities are defined in terms of two important criteria, namely, location and shared attributes. Their location either inside or adjacent to TFCAs means that the majority of citizens of participating countries cannot directly stake their claims in TFCAs. The results of the locational criterion are that there are logically very few residents who can directly participate in the affairs of cross-border nature conservation projects. It is generally believed that the population in the countryside, where TFCAs are created, is low.

The assumed and real common attributes of local communities are intriguing not so much because they hark back to the view of a community as a homogeneous group, but because they suggest that local communities should have clear social relations of some sort. In the context of southern Africa, the social relations of those communities mainly refer to linguistic or cultural groups among the black population. This is evident from references to the Nama in the Ai-Ais/Richtersveld Tranfrontier Park (ATP), the Shangaan in the GLTP and the Sotho in parts of the Maloti-Drakensberg Transfrontier Project (MDTP). In the last five years or so, consultants and government officials have encouraged cross-border visits among members of these groups, partly as a way of building rapport with local communities that are directly affected by the creation of TFCAs (Ramutsindela, 2007). In both these cases, group members on the South African side are often used as ambassadors for TFCAs. Their ambassadorial roles can be ascribed to South Africa's active involvement in TFCAs and the country's favourable legal conditions that, for example, allow for land claims in national parks that form part of TFCAs. Against this backdrop, it can be argued that the concept of local communities in TFCAs either reaffirm the existence of social groups or promote the reconstitution of such groups. That is to say that the process of creating TFCAs is paralleled or accompanied by that of organising residents in TFCA sites as a group. The principle underpinning that process is that residents can only participate in TFCAs as a group but not as individuals. It could therefore be inferred that a combination of Western and traditionalist notions of African societies as made up of groups such as tribes and the like is being reproduced through notions of local communities.

The Entry Points of Local Communities into the TFCA Arena

The question that arises from the definition of local community in cross-border conservation referred to above is the way in which local residents become party to, or are marginalised from, those projects. Case studies in southern Africa reveal that local communities are incorporated into the design of TFCAs through community projects, most of which are concerned with wildlife and nature-based tourism ventures. For example, the creation of wildlife management areas (hereafter called WMAs) in Botswana and conservancies in Namibia and Swaziland facilitated the entry of local communities into TFCAs without having, first, to sell the TFCA idea to those communities. In Botswana, WMAs were created in the mid-1980s as buffer zones for national parks and nature reserves, and to function as migratory corridors for wildlife (Broekhuis, 1997). In addition to these functions, they are also used as a strategy for controlling the utilisation of wildlife outside protected areas (Hachileka, 2003; Mbaiwa, 2004), and as a base for rural development. In the view of the Government of

Botswana, WMAs provide a 'rural base for commercial and non-commercial activities [which] provide employment and income opportunities for rural dwellers' (Botswana, 1985; see also Atlhopneng & Mulale, Chapter 8 and Moswete *et al.*, Chapter 11 in this book). In view of the extensive literature of CBNRM in Botswana, there is no need to describe them in detail here. They are only referred to in this chapter to demonstrate how the creation of WMAs brought local communities closer to the TFCA projects.

It should be noted that although Botswana was not a principal target of TFCAs, the first TFCA in post-1994 southern Africa, the Kgalagadi Transfrontier Park (KTP), was officially established on the Botswana–South Africa border in May 2000. The creation of the KTP was facilitated by, among other things, the location of the Gemsbok National Park (Botswana) and the Kalahari Gemsbok National Park (South Africa) back-to-back, and the absence of a fence or a strip of land separating them (see Ramutsindela, 2004). At the time of the establishment of the KTP in 2000, WMAs were already in place in Botswana. However, the creation and spread of TFCAs in southern Africa meant that WMAs in Botswana could be realigned with the TFCAs in one way or the other. Subsequently, attempts were made to create a corridor that links the TFP through the highly contested Central Kalahari Game Reserve to Chobe in the north of the country. The planned corridor is in line with an earlier map of TFCAs that was published by Getaway Magazine in 1996. The point here is that the corridor aims at linking the existing TFCA (i.e. the KTP) with nature reserves and WMAs in order to create an extensive TFCA across Botswana. From this brief summary of WMAs and TFCAs, it can be argued that their merging means that local communities for which WMAs were created are incorporated into the design for TFCAs. It can also be suggested that the creation of WMAs in future in Botswana, and wherever TFCAs are desirable, is most likely to be in concert with these cross-border schemes. Evidence of this trend is found in the creation of conservancies in Namibia and Swaziland.

Communal conservancies in Namibia are a recent phenomenon. Their emergence in postcolonial Namibia is ascribed to the promulgation of the Nature Conservation Amendment Act of 1996, which gave farmers in communal land rights in wildlife. Before this Act, the rights to use and benefit from wildlife were restricted to white farmers in privately owned land (see Scholz, Chapter 9 in this book). In light of this, the provisions of the Act of 1996 were lauded as progressive (Blackie, 1999). Communal conservancies are important for TFCAs in two main ways: their geography and overemphasis on wildlife. Logically, communal conservancies in Namibia are preoccupied with wildlife because the Act of 1996, to which they owe their origin, places emphasis on wildlife rights. Local communities have used those rights to set aside pieces of land for wildlife as part of the solution to people–wildlife conflict, and for commercial reasons. As most of the conservancies are located adjacent to existing protected areas, they contribute to the creation of corridors and buffer zones. More importantly, they bring communal land into areas earmarked for TFCAs while, at the same time, facilitating the merging of protected areas into TFCAs as the case of conservancies between Skeleton Coast and Etosha National Parks in the north-western part of the country demonstrates. In Swaziland, the creation of that country's first communal conservancy, Shewula, in 2002 provided the critical link needed for the creation of Lubombo TFCAs between Mozambique, South Africa and Swaziland.

It could be suggested that local communities on the South African side entered the TFCA arena through the legal route of land rights as the cases of the San and Richtersveld communities demonstrate. The San in and around the KTP successfully claimed their land in the Kalahari Gemsbok National Park in 1999, the same year that the governments of Botswana and South Africa signed an agreement for the establishment of the trans-frontier park. The restoration of their land rights through the constitutionally sanctioned land restitution meant that the San could not be completely ignored from the TFCA process, not least because they had recourse to the laws of the country. The same applies to the Richtersveld community on whose land the Richtersveld National Park, which forms part of the ATP, was established in 1991. In both cases, emphasis is placed on local communities on the South African side, in part because of the prominence of land right issues in South Africa. For example, the Treaty of ATP specifically names the Richtersveld community in South Africa as the 'local community' despite the fact that their fellow Nama people are located on the Namibian side of that TFCA as well. An important point that can be drawn from this experience is that the TFCA process gives more recognition to local residents whose land forms part of TFCAs more than those whose land does not. Further evidence of this is found in the description of the geographical area of the GLTP in the Treaty. The community on the Zimbabwean side is located in the area (the Sengwe corridor), which is the only link between Gonarezhou National Park (Zimbabwe) and Kruger National Park (South Africa). Against this backdrop, the Sengwe community has specifically been referred to in the Treaty. These cases show that land rights (in the formal or informal sense) provided avenues through which communities could be incorporated into the TFCA process.

Whether in the exercise of land rights or in the establishment of WMAs and conservancies, local residents have to participate as a group. The formation of such groups is not necessarily organic, but is guided by legislative requirements that should be met by people forming such groups. For example, communities claiming restoration of their land rights in South Africa are not only formed on the basis of a shared history of land disposses-sion, but are, more importantly, recognised through structures such as

Communal Property Association, which have a legal backing (see also Govender van Wyk & Wilson, Chapter 13 in this book). The Communal Property Association Act of 1996 recognises the community as a group of people who wish to have rights to or in a particular territory (South Africa, 1996). The Act creates the platform for reconstituting communities on historical claims to land or shared interest. Where such a shared history of land dispossession is not recognised as in WMAs (Botswana) and communal conservancies (Namibia), the legislative framework requires that clear boundaries around a particular group of people be demarcated. The official recognition of communal conservancy status in Namibia is dependent upon, among other things, a defined membership and clear physical boundaries of the conservancy (Shumba, 1998). This and other group formations play a pivotal role in the ways in which local communities access benefits from TFCAs.

TFCAs and the Beneficiation Framework

There are wild imaginations among officials, consultants and local communities about the benefits of TFCAs to the locals. These range from the reconstitution of pre-colonial social groups to the rebuilding of national economies, especially those of countries such as Angola and Mozambique that had seriously been disrupted by civil wars (Namibian, 2003). The root of the imaginations can be traced to the preambles to the treaties and the marketing of TFCAs. For example, the preamble to the Treaty of the ATP suggests that there is a conscious need to integrate cultural traditions, traditional land use and biodiversity conservation. However, in reality, land use is restricted to ecotourism, which is seen by officials and local communities as the main activity from which countries and their citizens stand to benefit from TFCAs. Indeed, the development of trans-border nature-based tourism is one of the primary objectives of TFCAs. Table 10.1 shows ecotourism estimates in selected TFCAs in southern Africa. It is claimed that nature-based tourism in the GLTP generates approximately $US2.45 billion per year (Spenceley, 2006). Ecotourism is not only referred to in the treaties, but is also impressed upon participating governments and their citizens from the inception of TFCAs. For example, in signing the Memorandum of Understanding for the establishment of Iona/Skeleton TFCA on the Angola–Namibia border, the Namibian Environment and Tourism Minister Philemon Malima is reported to have said that the TFCA 'will improve the conservation status of the northern Namib Desert as well as create opportunities for the people of the area' (Namibian, 2003, 5). This begs the questions of how local communities access those opportunities and the form in which those opportunities exist.

As can be gleaned from the discussion above, the ecotourism beneficiation framework is aligned with the place of local communities in TFCAs.

Table 10.1   Tourism potential and arrivals in TFCAs

Country	Conservation area	Carrying capacity and ecotourism potential (persons p.a)	Number of visitors per park in 2001	TFCA
Malawi	Kasungu	4700	1709	
	Nyika	31,300	1774	Nyika
	Luanga Valley	1,97,301	1709	
Mozambique	Coutoda 16	3,75,229	NA	GLTFP
	Zinare			
	Banhine, Limpopo Valley			
	Lago Niassa	1,56,085	NA	Selous-Niassa
	Futi corridor/ Maputo elephant reserve	18,096	200	Lubombo
Namibia	Ai-Ais Hot Springs	7,33,433	59,000	Ai-Ais/ Richtersveld
	Skeleton Coastal Park			
	Nat.West Coast Tourism Area			
	Etosha			
	Namib Naukuluft			
	Eastern Caprivi	1,02,736	NA	Okavango Upper Zambezi
	Etosha			
Swaziland	Lubombo conservancy	3570	NA	Lubombo
	Malolotja Game reserve	1600	NA	
Zimbabwe	Gonarezhou	1,31,530	5000	GLTFP
	Chimanimani	4840	4218	Chimanimani

(Continued)

Table 10.1 *Continued*

Country	Conservation area	Carrying capacity and ecotourism potential (persons p.a)	Number of visitors per park in 2001	TFCA
	Hwangwe (linked Victoria falls)	5,16,959	20,810	Okavango Upper Zambezi
	Mana Pools NP	1,09,751	9886	Luangwa Valley
	Hurungwe/ Chewore			
	Community land	NA	NA	Shashe Limpopo
South Africa	Richtersveld	61,080	5264	Ai-Ais/ Richtersveld
	Kruger (linked community Sabi, Timbavati)	9,98,358	9,33,672	GLTFP
	Kalahari Gemsbok	1,92,060	28,200	Kgalagadi
	Ndumo	66,345	4509	Lubombo
	Temba	NA	1618	
Botswana	Kgagalagadi	1,442,156	1110	Kgaladi
	Okavango, Chobe and Kasane		88,619	Okavango Upper Zambezi
	Central Kalahari		NA	
	Moremi, GR and Savuti NP		31,073	
Angola	Iona NP	41,927	NA	Iona/Skeleton Coast
	Luiana and Mavinga	1,05,470	NA	Okavango Upper Zambezi

Source: Blueprint Strategy and Policy (2002)

Those communities who own land in cross-border conservation areas have a voice in ecotourism initiatives found in those areas. After all, such communities agree in principle that their land will mainly be used for nature-based tourism. This is possible since local communities use their land as contractual parks in some of the transfrontier parks. In other words, communities engage in ecotourism in TFCAs in several ways, depending on the nature of their land rights or relationship with cross-border conservation projects. For instance, it is possible for local communities with land rights to enter into partnerships with the private sector to build lodges in TFCA areas from which they can collect revenue. Those partnerships could be in the form of concessions as the case of the GLTP demonstrates. Concession contracts are granted to the private sector for a specific period of time during which investors should recover their infra-structural costs and make profit. The principle underlying concessionaires is that the private sector invests to ecotourism with the hope that local communities will get jobs and also receive cash from fees and revenue generated from the enterprise. When the concession contract lapses, the ownership of the tourism infrastructure is ceded to the authority in charge of the area in question. Concessions have been granted on both the South African and Mozambican sides of the park (Box 10.1). While those on the South African side are largely ascribed to the commercialisation process that took place in 2000 (see also Mabunda & Wilson, Chapter 7 in this book), concessionaires in Mozambique are directly linked to the creation of the GLTP. At the beginning of 2007, the concessionaires were invited to tender for the development of three- to five-star lodges (a maximum of 96 beds) in Madonse Concession area located 10 km from the Giriyondo Entrance Gate, and to develop camp sites and self-catering units in Massingir Concession, which is around Massingir Dam (Business Times, 2007).

Box 10.1 'When Cultural Values are Marginalised': Impacts of the GLTP to Local People in Mozambique

ADERITO MACHAVA

Nature conservation measures have significant impacts on local people's lives but relatively little attention has been given to human consequences of conservation initiatives and related processes. When attention is given, the focus is typically a materialist: an eco-nomic analysis of what kinds of impacts certain initiatives have on local livelihoods. Economic welfare is undeniably important but still only one facet of local people's lived experience.

Studies on the impact of development-induced displacement have shown that the number of people affected (e.g. displaced) by programmes promoting national, regional and local development is substantial and usually they do not benefit from such development. Instead, they are more often impoverished, losing their cultural, economic and social resources and quite often provoking resistance.

This case example aims to discuss the local people's perception of *land* and *home*'s dimensions at Massingir District – Gaza province in Mozambique affected by the creation of GLTP. My aim is to argue that material and ideational perspectives are obviously intertwined. Therefore, there is the oft-expressed perspective that indigenous people and their communities and other local communities have a vital role in environmental management and development because of their knowledge and traditional practices.

Background and Theoretical Considerations

The GLTP was created in 2001 including Mozambique's Limpopo National Park, Zimbabwe's Gonarezhou National Park and South Africa's Kruger National Park. It is the largest initiative on the continent, which combines conservation, environmental protection, tourism and economic development.

Ongoing discussions have revealed that Induced-Development projects often have negative impacts on people unless good planning is made in consultation with the targeted communities. Historically, the majority of native populations around or within the parks have been disempowered, dislocated and relocated on the edge of parks boundaries or entirely outside the parks on soils less productive (Grimé, 2005).

Grimé (2005) points out that the initiatives behind conservation efforts around parks sound good on paper but often have negative consequences for local people, although she stresses that the intentions of conservation effort are admirable. Nevertheless, failure in addressing the needs within this framework is serious. Additionally, the policy creators often fail to understand the local population's viewpoint. Parks like this take ecological considerations into account first, leaving human populations as an afterthought most of the time. Grimé (2005) points out that these parks are in many cases created based on a 'racist, western, white viewpoint, where the importance of nature forces deprivation of resources by the local population'.

It is widely acknowledged that development-induced displacement is crisis-prone, even when necessary as part of broad and beneficial development programmes. According to Cernea (1995), it dismantles existing modes of production, disrupts social networks, causes the impoverishment of many of those uprooted, threatens their cultural identity and so on, increasing, as Koening (2002) mentions, vulnerability and dependence.

Planning and Consultation of GLTP

Some authors have shown the importance of good planning with participatory approaches (Cernea, 1995, 1999). However, studies have indicated the problems in the GLTP planning process (University of the Witwatersrand – Refugee Research Programme, 2002: 3). For example until 2003 it was not clear what would be done with the local people and communities concerned and how it would ensure an improvement in the socio-economic standing of the communities if they would be removed. This situation resulted from lack of consultation and planning with the different sectors of the population difficulties in bringing together all the parts concerned to the planning process.

From the very beginning, local communities that would be affected by the GLTP creation were not involved or they were unclear what would happen to them. Refugee Research Programme's study (University of the Witwatersrand – Refugee Research Programme, 2002: 7) points out that there was a lack of information about the Park itself at the grassroots level and there was a 'great deal of confusion among the communities in part due to contradictory information spread by different sources of information, including the Mozambican local government officials', and in the meantime, they have not been consulted about their views and options in relation to it.

My interviews with the local people in 2004 confirmed the Refugee Research Programme's findings:

> What I know about the park is that people have to move and I heard it from my relatives who said people have to move from this place. (Interview with Algina Valoi, Massingir, 29.04.2004)

Another interviewee, a migrant worker, said:

> I only heard about the park while in RSA. I have never heard about the park here in Massingir. (Interview with Almon Maluleque, Massingir, 29.09.2004)

It seems Mozambican authorities' failed to involve local communities in the development planning of 'Coutada 16'. This has cost the promoters much needed public support. People have also been critical and fearful over the prospect of large game animals living dangerously close to villages, crops, livestock and water sources.

Understandings of Place and Land

Traditionally, the land and the cultural and spiritual connections to it have been important for the local communities and their identities. According to Nkosi (2006):

> The sacredness of land in Africa is linked to the fact that ancestors are buried in it: Without land, we would not have a home for a dead body. That is why we kneel barefooted next to the grave when we want to communicate anything to our ancestors, showing a lot of respect for the land on which they lie. When death strikes in a family, no one is allowed to till the land. We mourn until that person is buried. After a funeral, in some cultures, we do not touch the soil with a hoe, do not plough or till the land until a ritual of cleansing the family is performed.

This is also evident but not recognised in the GLTP planning and creation process. In relation to the prospect of leaving their ancestral land, some of my respondents made similar connections as described by Nkosi (2006):

> I will not leave this place for them to bring animals. This place belonged to my forefathers and their remains are buried here. What do you think I can do? I will rather fight the animals than leaving this place (...) This Park only came to add more suffering on the people not to improve our livelihoods. (Interview with Derito Mate, Massingir 29.09.2004)

As the Refugee Research Programme's study (University of the Witwatersrand – Refugee Research Programme, 2002: 11) mentions the 'focus on accessing resources does not take into account that there is an inherent value in the land where people reside'. Indeed, people revealed a strong attachment to the villages and places where they were born, to the land where their ancestors farmed and were buried, and to sacred trees and idols.

To conclude, this case study advocates that there is a need of an improved capacity for management of conflicts between conservationists and local communities through honest and sincere dialogue. In this respect governmental bodies should recognise and duly support their identity, culture and interests and enable their effective participation in the achievement of sustainable development. The frameworks adopted in addressing the negative impacts of displacement caused by induced-development projects must take into account historical, anthropological, sociological and ethnological data. It must be inclusive, bringing together all interested groups (e.g. ethnic). This is the possible way forward to build comprehensive and informed conflict management frameworks that can be more effective.

The risks involved in such ventures are that local community interests could be overtaken by the private sector, and that the whole idea of poverty alleviation could flounder. For example, the San in the KTP protested outside the South African Parliament in 2004 to demand government intervention into the use of their land in the TFCA. They were poor and unhappy that they had no control over their land that was being used by professional hunters, as the Khomani San leader David Kruiper commented: 'my heart is very heavy today, because we have no control over the land that the President [Mbeki] gave to us and I want him to tell us what we should do' (cited in Cape Argus, 2004: 3). The concessionaires could operate below profit as is the case on the South African side of the GLTP (Spenceley, 2006).

Local communities participating in WMAs and conservancies adjacent to TFCAs often do not have direct access to TFCAs. Consequently, their lodges and other enterprises are located outside the formal boundaries of transfrontier parks. For example, in WMAs in Botswana local communities derive cash from controlled hunting areas (CHAs), which are managed through community or development trusts (see Mbaiwa & Darkoh, Chapter 12 in this book). For example, the Okavango Community Trust generated $UD 1 500 000 in 2001 (Mbaiwa, 2004). The concept of community referred to above implies that such revenue benefits villages within the boundaries of the 'community areas'. Conservancies operate along similar lines. What is instructive about CBNRM areas such as WMAs and conservancies is that the benefits they derive or access are not directly linked to TFCAs, although their areas feature in the grand plans for those cross-border conservation projects. The chance for these versions of CBNRM to tap into the tourism market rests in their proximity to TFCAs. This is not to suggest that CBNRM projects located away from TFCAs cannot access the tourism market. Rather, it is argued, consistent with the theme of this chapter, that CBNRM benefits are peripheral to the TFCAs beneficiation framework.

Concluding Remarks

The advent of TFCAs has refocused the links between society and the environment at a higher scale than that of CBNRM, mainly because of the sphere of influence of TFCAs and their real and expected impact on the conservation of biodiversity and socio-economic development. Whereas the environmental focus of TFCAs is the protection and management of ecological systems, the socio-economic rationales for TFCAs coalesce towards local economic development through nature-based tourism. The minimum requirements for achieving these goals are that habitats should be reorganised into bioregions while humans in or around them (i.e. bioregions) form manageable groups. This chapter has shown that the bioregional logic underpins the creation of TFCAs, hence the use of protected areas as their core. In turn, protected areas give TFCAs the necessary legal backing. The chapter also established that the concept of community is strongly entrenched in the discourse of cross-border nature conservation. The concept does not only provide the basis on which local residents are incorporated into TFCAs but also leads to a differentiated access to opportunities in and around those conservation areas. Categories of communities that stand out are those with land rights inside TFCAs and those using or managing natural resources in areas outside but adjacent to TFCAs. On both counts, ecotourism is considered the catalyst for economic benefits. Finally, the discussion in this chapter calls for a deeper understanding of the nature of the relationships between local communities and TFCAs, and the outcomes of those relationships beyond the statistical assessment of revenue from ecotourism in TFCAs.

References

Adams, W. and Hulme, D. (2001) Conservation and community. In D. Hulme and M. Murphree (eds) *African Wildlife and Livelihoods* (pp. 9–23). Oxford: James Currey.

Bennett, A.F. (1990) Habitat corridors and the conservation of small mammals in a fragmented forest. *Landscape Ecology* 4, 109–122.

Bennett, A.F. (2003) *Linkages in the Landscape: The Role of Corridors and Connectivity in Wildlife Conservation*. Gland: IUCN.

Blackie, R. (1999) *Communities and Natural Resources: Trends in Equitable and Efficient Use* (Research Discussion Paper 29). Ministry of Environment and Tourism: Windhoek.

Blueprint Strategy and Policy (2002) *Transfrontier Conservation Areas: Tourism*. (Assessment Final Report). Stellenbosch.

Botswana (1985) *National Development Plan*. Gaborone: Government Printer.

Brandon, K., Redford, K.H. and Sanderson, S.E. (eds) (1998) *Parks in Peril: People, Politics and Protected Areas*. Washington, DC: Island Books.

Broekhuis, J.F. (1997) Land use planning for wildlife conservation and economic development. In *Proceedings of a National Conference on Conservation and Management of Wildlife in Botswana* (pp. 10–151). Gaborone.

Brunckhorst, D.J. (2002) *Bioregional Planning*. London: Routledge.

Business Times (2007) Request for pre-qualification. Concessionaires for tourism facilities in Parque Nacional Do Limpopo, Mozambique (20 January, p. 9). Johannesburg.

Cernea, M. (1995) Social integration and population displacement. *International Social Science Journal* 143 (1), 91–112.

Cernea, M. (1999) Why economic analysis is essential to resettlement: A sociologist's view. In M. Cernea (ed.) *The Economics of Involuntary Resettlement: Questions and Challenges*. Washington, DC: World Bank.

Cape Argus (2004) Intervention: David Kruiper consults with Independent Democrats leader Patricia de Lille (19 August, p. 3). Cape Town.

Grimé, N. (2005) Anthropological perspectives of transboundary park impact: People of the great limpopo transfrontier Park, Southern Africa. In D. Harmon, (ed.) *People, Places, and Parks: Proceedings of the 2005 George Wright Society Conference on Parks, Protected Areas, and Cultural Sites.* Hancock, Michigan: The George Wright Society.

Hachileka, E. (2003) Sustainability of wildlife utilisation in the Chobe District, Botswana. *South African Geographical Journal* 85, 50–57.

Hilty, J.A., Lidicker, W.Z. and Merenlender, A.M. (2006) *Corridor Ecology: The Science and Practice of Linking Landscapes for Biodiversity Conservation.* Washington, DC: Island Press.

Hutton, J., Adams, W.M. and Murombedzi, J.C. (2005) Back to the barriers? Changing narratives in biodiversity conservation. *Forum for Development Studies* 2 (2), 341–370.

Koening, D. (2002) *Toward Local Development and Mitigating Impoverishment in Development-Induced Displacement and Resettlement (Working Paper Series).* Oxford: Refugee Studies Centre.

Mbaiwa, J.E. (2004) The success and sustainability of community-based natural resource management in the Okavango Delta, Botswana. *South African Geographical Journal* 86 (1), 44–53.

McNeely, J.A. and Pitt, D. (eds) (1985) *Culture and Conservation: The Human Dimension in Environmental Planning.* London: Croom Helm.

Namibia (2003) Namibia and Angola embrace conservation beyond borders (5 August, p. 5). Windhoek.

Nkosi, Z. (2006) Spirituality, land and land reform in South Africa. On www at http://www.wcc-coe.org/wcc/what/jpc/echoes-16–05.html. Assessed 10.2.2008.

Ramutsindela, M. (2004) Glocalisation and nature conservation strategies in 21st century Southern Africa. *Tijdscrif voor Economische en Sociale Geografie* 95, 61–72.

Ramutsindela, M. (2007) *Transfrontier Conservation Areas in Africa: At the Confluence of Capital, Politics and Nature.* Wallingford: CABI.

Shumba, M. (1998) Devolving of natural resource tenure in Namibia. In E. Rehoy (ed.) *Natural Resource Tenure in Southern Africa: Exploring Options and Opportunities* (pp. 74–82). Gaborone: SADC Secretariat.

South Africa (1996) *Communal Property Association Act.* Pretoria: Government Printer.

Spenceley, A. (2006) Tourism in the Great Limpopo Transfrontier Park. *Development Southern Africa* 23 (5), 649–667.

Terborgh, J. (1999) *Requiem for Nature.* Washington, DC: Island Press.

Treaty of the Ai-Ais/Richtersveld Transfrontier Park (2003) *Treaty Between the Government of the Republic of South Africa and the Government of the Republic of Namibia on the Establishment of Ai-Ais/Richtersveld Transfrontier Park.* Pretoria: Department of Environmental Affairs and Tourism.

University of the Witwatersrand – Refugee Research Programme (2002) *A Park for People? Great Limpopo Transfrontier Park – Community Consultation in Coutada 16,* Mozambique.

Zbicz, D.C. and Green, M.J.B. (1997) Status of the world's transfrontier protected areas. *Parks* 7, 5–10.

Interviews (Box 10.1).
Interview with Almon Maluleque, Massingir, 29.09.2004.
Interview with Derito Mate, Massingir 29.09.2004.

Part 3

Tourism, Local Communities and Natural Resources in Transition

Chapter 11

Village-based Tourism and Community Participation: A Case Study of the Matsheng Villages in Southwest Botswana

NAOMI MOSWETE, BRIJESH THAPA and GARY LACEY

Introduction

Village-based tourism has grown in popularity among international tourists largely due to the availability of diverse cultural and natural resources. This type of tourism is a subset of cultural tourism, in which visitors have an opportunity to observe and participate in daily activities, customs and traditions of local people. Also, food and services are locally produced, and tourism revenues generally stay within the local communities. Village tourism includes all activities that range from locals organizing tours of cultural heritage sites, to sightseeing within the villages (Burns & Barrie, 2005; Frost, 2003; Thai, 2002; Zeppel, 2002). In addition to economic benefits, it also promotes conservation of both the natural and cultural environments with minimal impacts (Dei, 2000; Intrepid Travel and Victoria University, 2002; Rattanasuwongchai, 1998). Village-based tourism is often promoted as a panacea for economic development in poorer regions of both developing and developed countries such as Nepal (Nepal, 2005), China (Nyaupane *et al.*, 2006; Ying & Zhou, 2007), Taiwan (Hong, 1998), Japan (Moon, 1989), Australia (Dyer *et al.*, 2003) and England (Holloway, 2007).

In the context of Africa, village tourism has been identified as an important regional development and poverty alleviation tool. Throughout the region, tourism is responsible for the creation of the greatest proportion of jobs and socio-economic benefits for rural dwellers (Ashley & Jones, 2001; Barnes, 2001; Burns & Barrie, 2005; Dieke, 2000; Roberts & Hall, 2001). More specifically, the southern African countries have been successful in village-based tourism due to the abundance of cultural heritage, wildlife and natural resources that are promoted via the Community-based Natural

189

Resources Management (CBNRM) programme. Within this programme, wildlife and village tourism resources are controlled by local communities, who are able to capitalize and accrue direct economic benefits (Box 11.1). The main goal of the CBNRM is to create a 'foundation for conservation-based development, in which the need to protect biodiversity and ecosystems is balanced with the need to improve rural livelihoods and reduce poverty' (GOB, 2005: 5).

Box 11.1 Communal Areas Management Programme for Indigenous Resources in Zimbabwe (CAMPFIRE)

HARETSEBE MANWA

The Zimbabwe government has set aside 13% of its land as the Parks and Wildlife Estate for nature and conservation. Approximately 2.1% is forest reserve. A total of 37% (18.749 m²) of the parks and wildlife estate has been set aside as safari areas (Ngwarai, 2000). In addition to the parks and forest reserves, approximately 17% of land has been taken from both commercial and communal lands. This is the land under Conservancies and CAMPFIRE projects, respectively. The chapter focuses on challenges in implementing the CAMPFIRE programme.

Community Involvement

Studies on impacts of tourism have confirmed the importance of involving local communities at various stages of tourism development from planning through implementation to its evaluation (Murphy, 1985). Failure to involve local communities has resulted in resentment and negative social and environmental impacts (Sofield, 1991).

The success of the community-based sustainable resource management is based on a number of assumptions:

(1) that there is decentralization of decision-making to the community;
(2) that the community acquires managerial skills;
(3) that the user group is homogeneous and small; and
(4) that the benefits derived from resource management should benefit communities mostly affected by their proximity to the resource or because they have to change their lifestyle to accommodate wildlife.

A good example is of the San people who were moved from their traditional habitat and settled among the Tonga people to make way for the creation of community-based wildlife parks (Mberengwa, 2000). The Tonga people could be said not to have suffered as much as a result of the relocation as the San people.

Decentralization of Decision-making

The first assumption of community-based sustainable resource management is that it would decentralize decision-making to local communities. This was perceived as an effective way of democratizing decision-making and strengthening democracy at grassroot level thereby empowering ordinary people to have a greater say over development programmes in their own areas as well as the facilitation of the interaction with government agencies (Conyers, 1990). In practice, however, the process is often more top down whereby decision-making is more central government – directed and controlled through the Rural District Councils (RDCs) (Campbell *et al.*, 1999; Makumbe, 1998; Virtanen, 2003; Wekwete and Mlalazi, 1990). Local people have been reduced to receivers of government handouts (Makumbe, 1996). This is a result of the domination of RDCs by representatives of central government who influence local activities (Mutizwa-Mangiza, 1990). Communities who bear the brunt of the costs of crops raided by animals do not yet have a legal right to accrue revenue themselves. Negotiations are normally held by the RDCs who are also responsible for deciding on the disbursements of proceeds. Although the proceeds should be shared equally between the RDCs and local communities, often RDCs allocate all the money to cove their operational costs, for example, salaries and wages of council employees (Hasler, 1996).

Acquisition of Managerial Skills

Another importance of development projects is that they help to equip local communities with managerial skills to run development projects (Cottrell, 2001). In Zimbabwe, however, most donor-funded projects employ external experts to manage development projects. There is no mechanism in place to facilitate eventual takeover by local communities through either internship or understudy of current management. When donors eventually pull out, the projects suffer because communities lack core skills to sustain a project, for example, marketing functions, management functions, record keeping and even knowledge of who their customers are.

Homogeneity of the Group

The second implicit assumption under the CAMPFIRE programme is that the rural poor are homogeneous. Evidence has however shown that even among the rural poor there is a stratification into poor and the very poor. Rondinelli (1993: 70) referred to the very poor as the 'special publics'. These are people who lack the political muscle, uneducated and too poor to have their voices heard. Consequently, the majority overshadows them. Similar sentiments are echoed by Joppe (1996), who has questioned the whole concept of community development as being nothing more than a perpetuation of past injustices. In the CAMPFIRE projects, social class is also used to determine who should represent villagers in decision-making committees, for example, the RDCs. The nature of such committees requires someone who is articulate and has at least a High School Diploma. This tends to favour the local elites for example, the local school teachers, or business people (Hasler, 1996).

Allocation of Benefits

The third assumption of the CAMPFIRE programme is that resource management should ensure that the people who should benefit the most are those whose lives have been changed drastically by the introduction of the programme (Hasler, 1998; Motavalli, 2002). In practice, however, the introduction of the CAMPFIRE in some cases has benefited the larger community and not the intended beneficiaries (Madzudzo, 1998).

Because benefits to local communities are so minimal, local communities feel that the government instead of just policing people against killing wild animals could provide them with skills and implements to improve their yields (Gumbo, 1992).

Most of the CAMPFIRE communities' livelihoods partly hinge on hunting, fishing and partly on use of other resources. The Zimbabwe government regards recreational sport hunting as an economically and ecologically efficient use of wildlife consistent with a policy of high-quality and low-density tourism (Government of Zimbabwe, 1992). The Zambezi River where most of the CAMPFIRE communities live is endowed with over 70 species of fish, including bream, tiger fish and other small varieties that have become important for commercial and sport fishing (Hoole, 2001). The total revenue earned from sport hunting in safari areas in 1999 ranged from Z$87.5 million to about Z$110 million (US$1 = Z$28) (Ngwarai, 2000).

There are however, restrictive measures on who can engage in hunting. The first restrictive measure is the high cost of engaging in such an activity, which is unaffordable for most Zimbabweans who are hard pressed by current economic problems such as hyperinflation that was estimated to be around 2000 (Central Statistics Office, August, 2007).

The second exclusionary measures are regulation governing commercial hunting in Zimbabwe. Instead of taking advantage of traditional methods of hunting which local communities have practiced for generations, the regulations only recognize passing of a practical and written test and registration with the Parks and Wildlife Management Authority in Zimbabwe. This requirement further deprives communities of their livelihood (Mberengwa, 2000).

Conclusions

The CAMPFIRE has been acclaimed as a model for other countries to follow in order to conserve the natural resources because of the benefits alluded to in this chapter. The chapter has tried to demonstrate that the model faces a lot of challenges, but there are merits to the model if these challenges are addressed. What is important is that the Zimbabwe government could use the model in allocating land in the conservancies and wildlife areas. The success of CAMPFIRE in the resettlement area will only succeed if the following conditions are met. First, there needs to be education of the resettled communities on the importance of the natural resources. This should include how these resources can be used sustainably instead of the current utilization of the land for crop farming (especially if the area is unsuitable for intensive crop farming). The second intervention would be mentoring of the resettled communities by attaching village leaders to successful CAMPFIRE projects to gain a first-hand experience on how the programmes are managed.

Successful ventures in CBNRM are evident in numerous countries within southern Africa (Dei, 2000; Taylor, 2007). Among the countries in the southern African region, Botswana has advocated sustainable economic use of wildlife through CBNRM, especially in the designated Wildlife Management Areas (WMAs) (GOB, 1986, 1992; DWNP, 1999). Wildlife is central to the lives and livelihood of rural dwellers in Botswana. Rural people are still dependent on this resource for subsistence hunting as well as for both consumptive and non-consumptive forms of tourism (Barnes, 2001; Granlund, 2001). The government has several wildlife conservation policies that are intended to help local communities (GOB, 2002). The Community-based Natural Resources Policy is designed to devolve management of wildlife, land rights and other natural and cultural resource management and ownership to rural communities. Such a policy of decentralization of management of resources has been a major catalyst to promote socio-economic well-being of communities living in rural and impoverished areas (see also Becker and Scholz, Chapter 9 in this book). Under the policy, community-based organizations (CBOs) are responsible for the operations and management of resources.

CBOs are usually allocated a concession area and have acquired a certain security of tenure through a lease arrangement. The lease gives them exclusive rights over the wildlife quota and commercial tourism operations in their area. They can decide whether or not to hunt, and how to hunt the quota. Species are generally divided among the community members for subsistence hunting, or the quota can be sold to a private sector partner. Usually the organization sells the commercially valuable species such as elephant, zebra, lion and leopard to the private sector partner for safari hunting. Such trophy hunting joint venture agreements generate large sums of money for the community, and substantial employment during the six months of the hunting season (Gujardhur, 2001). In the non-hunting season (October–March), the hunting infrastructure is often used to accommodate photographic safaris, which are usually organized and managed by the joint venture partner.

Currently, more than 50 CBOs in various stages of development are involved in projects that range from thatching grass and herbal tea collection and marketing, to handicraft production, campsite management and trophy hunting joint venture agreements with the private sector. The majority of successful community-based projects are wildlife and wilderness related with trophy hunting joint venture agreements (Botswana Review, 2005). Community-based tourism initiatives are still relatively new in Botswana, and numerous projects are in the initial stages of development. The government denotes community-based tourism as 'tourism initiatives that are owned by one or more communities, or run as joint venture partnerships with the private sector with equitable community participation, as a means of using natural resources in a sustainable manner to improve their standard of living in an economic and viable way' (GOB, 2005: 6; see also Atlhopneng & Mulale, Chapter 8 in this book).

The government is committed to promoting tourism as a tool for economic development in rural communities. Some of the key tourism policy objectives are to (i) increase the number of citizens meaningfully involved in and benefiting from the tourism industry, (ii) decentralize management of wildlife tourism to rural areas and (iii) promote socio-economic well-being of the communities living adjacent to protected areas (GOB, 1990: 2). To meet these objectives, CBNRM schemes along with the operational arm of CBOs are being advocated and implemented in rural areas. Given that rural tourism ventures are still in their infancy, the objective of this research was to evaluate the role of a CBO and its tourism initiatives.

Tourism and Botswana

Botswana is a landlocked country located in southern Africa, and bordered by South Africa, Namibia, Zimbabwe and Zambia. It gained its independence from Britain in 1966. The country has a land area of about 582,000 km^2 with 1.7 million inhabitants at a low population density of 2.6 people per square kilometre living in largely rural areas. About 80% of the population is concentrated in the eastern region of the country where locals engage in pastoralism and agriculture in the fertile lands (GOB, 2001; WTTC, 2007).

The country has the greatest numbers and variety of wildlife in southern Africa, with the least crowded parks and game reserves (BTDP, 1998). About 40% of the land mass has been designated as protected areas, which include national parks, game reserves and WMAs (Botswana Review, 2005). Some of the popular protected areas that have experienced substantial increases in visitation are the Kgalagadi Transboundary Park, Central Kgalagadi Game Reserve, Chobe National Park, Moremi Game Reserve and Makgadikgadi National Park. Given that safari-based tourism and hunting are the most popular activities among visitors, wildlife and wilderness resources are major products that are strictly controlled.

Tourism is the second largest revenue earner after diamonds (BTDP, 1999). In 2005, over 1 million international visitors came to Botswana, which generated US$562 million (Botswana Review, 2005; WTO, 2005). The majority of visitors originated from the southern African countries, and the growth has been experienced at 13.7% per year during the last decade (BOPA, 2007; BTDP, 1999). Although future growth has been projected, the government's strategic plan is to maintain its focus on 'low-volume, high-price' activities, hence discouraging mass tourism in the country.

Tourism activity is largely concentrated in the northeast region of the country and is the major source of income and employment for residents. However, the southwest region is in dire need of economic development as 71% of people are estimated to be living in poverty, and 59% are described as very poor. The highest proportions of the latter class are remote area dwellers. People within this category are characterized by a lack of formal training and an inadequate education, compounded by weak institutions and poor leadership (Ditlhong, 1997; Phuthego & Chanda, 2003).

The southwest region of the country is very remote, and the economy is based on cattle rearing for meat production, with 90% of the export livestock sales going to the Botswana Meat Commission (Ministry of Local Government and Land (MLG), 2003–2009). In addition to livestock farming, the largest source of employment is from government and council jobs in the district (e.g. drought relief projects). In addition, the handicraft industry and hide-tanning provide a livelihood to some local communities. Tourism activity is extremely low and underutilized. However, national parks and reserves have begun to attract an increasing number of tourists with a potential for future growth, as the major attractions include unspoiled wilderness, a variety of wildlife and handicrafts of the San/Basarwa. Nevertheless, there are significant challenges to tourism development. Some of the challenges are a low awareness of tourism, poor roads to wildlife attractions, inadequate marketing of the region, unreliable water supplies, poor infrastructure for tourism, low levels of literacy among the population and remoteness of the region.

Plate 11.1   Entrance to KD1 (hunting zone) Wildlife Management Area (Photograph courtesy of Naomi Moswete, 2004)

In an effort to reduce poverty, promote self-reliance and encourage participation of local communities in national development activities, the Kgalagadi District Development Plan Five (DDP 5) was created. Based on a study, it was also determined that the tourism potential in the Kgalagadi District was largely untapped and that community hunting, managed through the CBNRM, could significantly increase the revenues generated from wildlife (MLG, 1997–2003). The Kgalagadi District Tourism Plan was created in 1996 to expand the economic base of the region. Recommendations were made for the establishment of a Community-based Development Trust (Johnson, 1996), which contributed to the creation of the CBNRM, and the subsequent formation of Nqwaa Khobee Xeya Trust entrance signpost (Plate 11.1) and Khawa Kopanelo Development Trust, both located in the district.

The creation of the Trust and numerous tourism-related initiatives has provided some level of economic activity. However, there are numerous challenges facing the residents of the Kgalagadi District with issues such as difficulty in accessing credit, low levels of entrepreneurial skills and rural–urban migration that has weakened the rural district's ability to leverage tourism. The purpose of this study was to examine village-based tourism initiatives among two settlements (Ukhwi and Ngwatle) in the Matsheng area of the Kgalagadi District. Three research questions were formulated:

(1) What type of village-based tourism initiatives exist in the Ngwatle and Ukhwi settlements?

(2) Are local people involved in the management of the activities and benefiting from it?

(3) What factors contribute to the success and/or failure of village-based tourism?

The Matsheng Area

The Matsheng Area is remote, located on the salt pan system of the Kalahari Desert in the district and about 550 km from the capital city. The region receives about 300 mm average annual rainfall that occurs erratically during the wet season (Granlund, 2001). In the summer, the average maximum temperature exceeds 41°C, with the winter minimum as low as –8°C (Chanda & Totolo, 2001). The area consists of four villages that are surrounded by WMAs and three small settlements. The communally managed wildlife rangelands were created in consultation with the local villagers for the protection and conservation of wildlife (GOB, 1986). In these areas, subsistence livestock and non-intensive, small-scale, arable farming are permitted along with controlled game harvesting but similar commercial activities are forbidden, as is the domestication of wild animals.

The three settlements consist of Ngwatle, Ukhwi and Ncaang, and are located within a controlled hunting zone known as KD1. Settlements are less formal and consist of a few household clusters that lack health clinics, schools and a regular water supply. They are usually inhabited by people of nomadic background. Generally, the total population in a settlement is above 150 but less than 500 inhabitants, with people mainly subsisting on hunting and gathering. However, villages are more formal establishments with populations of 500 or more that possess health clinics and schools.

The area has traditionally been associated with hunting and gathering activities of the aboriginal, San/Basarwa people. Currently, it is occupied by a more diverse ethnic mix, which includes the Bakgalagadi and Bangologa who originated in southwestern Botswana (Hitchcock, 1997; NKXT, 1999). The tribes do mix and the San/Basarwa, at one time, did hunt for the Bakgalagadi in exchange for domestic animals and food products but they have never integrated into a close knit single community (Hitchcock, 1997; NKXT, 1999).

The population of the KD1 area is less than 1000 people, with 454 living in Ukhwi and 206 in Ngwatle. Almost half the population is under 16 years of age (NKXT, 1999). Seventy percent of KD1's population is San/Basarwa. In Ukhwi, 67% are San/Basarwa and 33% Bakgalagadi, while Ngwatle is predominantly a San/Basarwa settlement (NKXT, 1999). Illiteracy rates are extremely high numbering approximately 75% of the adult population. Levels of literacy are at their lowest among the San/Basarwa, and only a small portion of the population of Ngwatle has any formal education (Chanda & Magole, 2001; NKXT, 1999). The livelihood practices of the settlements are based on the collection of veldt products, such as wild plants for herbal medicines and teas, forest berries, tubers and hunting. These practices are supplemented by labour remittances from the Local Government Drought Relief projects and other government welfare programmes (Chanda & Magole, 2001; Chanda *et al.*, 2003; MLG, 1997–2003). The exclusive reliance on hunting and gathering sets Ukhwi and Ngwatle apart from the other Matsheng villages, whose communities subsist primarily on livestock, small-scale crop farming and informal employment.

Ngwatle and Ukhwi are designated Remote Area Dweller (RAD) settlements and as such, are regarded as vulnerable. The government of Botswana has several policy initiatives to support income generation through wildlife utilization and rural tourism in designated RADs. The District Development Plan aims to generate small-scale tourism projects for the local community (DDP 4, 2003–2009), but a lack of supporting infrastructure and inaccessibility to potential tourism sites present major challenges for development. There are virtually no social services in Ngwatle. The only two buildings in the settlement are the drought relief structure that was built by the Kgalagadi District Council through Drought Relief Projects and a tannery that was built with revenue generated from the CBO (NKXT). There is one medium-sized government house that was built for use by the visiting, family nurse-practitioner from Ukhwi and/or Hukuntsi. The shelter is also used for storing monthly food supplies for the residents of Ngwatle. Of the three settlements in KD1, Ukhwi is the most developed. Social services and physical facilities in the settlement include a primary school, police station, cooperative shop, mobile health stop, a community guest house and craft shop and a borehole that produces saline water.

Overall, the communities have few opportunities to participate in the cash economy and are heavily reliant on the vagaries of the Kalahari's wet season. Given the national government's emphasis on the development of community-based rural tourism initiatives, the question arises as to the extent to which the Matsheng villages have benefited from tourism via the CBO Nqwaa Khobee Xeya Trust.

Case Study: Village-based Tourism in Matsheng Villages Methodology

A household list that contained all the information about settlements and villages from the Central Statistics office was obtained and used. Data were collected in June–July 2004. Primary data sources included household surveys and focus group discussions. Additional secondary data were derived from sources such as district development plans and maps, government reports, promotional brochures, policy documents and other published materials.

Of the total 94 households, 49 heads of the households were randomly selected in Ngwatle ($N = 16$) and Ukhwi ($N = 33$). Of the household heads, 67% were male and 33% were female. In Ngwatle, 75% were male, whereas in Ukhwi, 64% were male. Since the household surveys were conducted among heads of the households, they were overwhelmingly completed by males, which is consistent with Botswana tradition. For both areas, formal educational standards were very low. Since literacy levels in the two settlements were also very low, participants were asked closed, open-ended and pre-coded questions on issues of village-based tourism and demographic profiles. Questions assessed tourism awareness and importance, participation in community trust initiatives, general village tourism management and perceptions of tourism development in the study areas. In addition, a focus group interview was conducted in each of the settlements of Ngwatle and Ukhwi.

Major tourism attractions

Wildlife resources, including trophy animals such as lion, cheetah and leopard are found in relatively high densities in KD1. Other large animals found in abundance include various species of antelope, hartebeest, wildebeest and gemsbok. In addition, desert birdlife is plentiful too, including ostrich, social weavers and birds of prey (Roodt, 2004). The settlements of Ngwatle and Ukhwi are in proximity to the Kgalagadi Transfrontier Park and the Kutse Game Reserve. Wildlife tourism including safari and recreational hunting, photography and bush or wilderness camping are popular activities.

Besides the wildlife resources, the natural attractiveness of the study sites is enriched by splendid sand dunes, fossils and river valleys. Other natural attractions include the unique desert scenery, the salt pan system and the vast open desert landscape. In addition, the cultural landscape that includes the tangible and intangible culture of the San/Basarwa is an attraction to visitors. Others relate to historic resources and the cultural heritage, including the handicraft industry (Table 11.1).

Existing community tourism initiatives

Based on the CBNRM, a community development trust (Nqwaa Khobee Xeya Trust) was established in 1998 as the CBO for the three settlements. The Trust (NKXT) has been granted resource user rights over the controlled hunting area of KD1. The CBO has established a joint venture partnership with the safari operator that permits the company exclusive rights to conduct hunting and photographic safaris in the area. The operator has

Table 11.1   Tourism-related attractions

Type of resources	Descriptions	Level of use
Unspoilt salt pans	Mineral-rich pockets with high nutritional value grasses. These are vital wildlife habitats which are areas of great geological interest	High use. Photography; wedding functions by locals and tourists. Provide salt licks for livestock and wildlife
Sand dunes (ridges)	Elongated unique red soil ridges and mounds of different shapes and sizes, some are partially stabilized by a sparse vegetation cover, others are active and mobile	Low use. Used for ecotourism including expeditions, photography and filming
Fossil River valleys	Scenically attractive, grassy, shrubs, winding natural features. The valley floors are wildlife habitats. Large parts of it remain unspoilt	Minimal use. Sightseeing; wildlife viewing and photography and cultural activities
Multi-coloured pebbles found in pans	Shiny and unique colour found only in Ukhwi pan. Are of great geological interest	Low use. Used for photography and viewing by locals and visitors
Kalahari natural cultural landscape	Vast open bare land, with shrubs, undulating red soil ridges, depressions, valleys	Medium use. Photography, filming, sightseeing, research, wilderness camping
Forest (veldt) resources	Variety of edible forest food: herbs, roots, fruits, tsamma melons, devil's claw, moretlwa	High use. Social and cultural purposes, diet, medicinal, rituals
Cultural attractions	Appeal for locals & traditional cultures are high. Cultural products such as wooden and bone artefacts and beadwork is a popular activity	Moderate use. Generally, there is very low culture-based tourism activity in the area
Historical places of interest	Major villages on pans and valleys, San/ (Basarwa) settlements and sites (Middle to late Stone Age sites)	Minimal use: They are not known hence not used for tourism

been allowed to set up its luxury tented camps at developed campsites. Also, locals are given the opportunity to offer traditional dancing, demonstrations of traditional healing rituals, veldt product gathering and tasting and traditional hunting safaris. About 25 local jobs have been created including individuals who work for the safari operator as animal trekkers and skinners. In exchange for the exclusive rights, the community receives cash for the sale of 30% of the community quota, whereas 70% is subdivided among community households for subsistence hunting (focus group). The cash income is shared equally among the settlements. Carcasses (meat, skin and hides) from hunting remain the property of the community.

Other developments as a result of tourism are also evident in the area. Currently, small-scale handcraft production (e.g. ostrich egg shell products) is promoted for sale to visitors and for craft shops in Ghanzi and Kang. A proposed cultural centre is to be built in Ngwatle to exhibit and sell local crafts. A Trust office (with a project vehicle) and guesthouse in Ukhwi and a tannery shelter in Ngwatle have been completed (Plate 11.2). Six campsites were also constructed near Ukhwi and Ngwatle pans, some with long-drop pit-latrines and showers, fenced *bomas* (meeting places) and fireplaces (Plate 11.3). Projects were community driven as they selected and approved the locations and designs for the campsite infrastructure. The campsites have been spaced adjacent to the main access roads, the

Plate 11.2 Nqwaa Khobee Xeya Trust CBO guesthouse in Ukhwi (Photograph courtesy of Naomi Moswete, 2004)

Plate 11.3   Nqwaa Khobee Xeya Trust CBO campsite in Ngwatle (Photograph courtesy of Naomi Moswete, 2004)

Kgalagadi Transfrontier Park and 4 × 4 trails. All projects were built by the local people with the assistance of an international NGO.

Knowledge and importance of tourism

Respondents were assessed of their knowledge and their views pertaining to the importance of tourism in their area. About 64% in Ukhwi and 75% in Ngwatle indicated that they had no knowledge of tourism. Upon further probing, 94% of respondents in Ukhwi and 87% in Ngwatle reported that tourists visited their village. Many of the respondents opined that tourism was about foreigners coming to their village, buying craftwork and taking photographs. Basically, tourism was perceived as an opportunity to sell products to foreigners. Tourist activities were perceived as taking photographs, buying local products and visiting parks and farms. Similar sentiments were also echoed in focus group interviews as the majority did not understand that tourism was an industry that locals could initiate and participate in for their own benefit. However, 94% of respondents in Ngwatle and 88% of those in Ukhwi noted that tourism was important in the KD1 area. Also, a high proportion of respondents was in favour of more tourism activities, and believed that increased tourism would bring about development and greater employment opportunities. Overall, more respondents from Ukhwi than Ngwatle indicated support for tourism promotion and activities.

Resident participation and involvement

Given the existing tourism initiatives in Ngwatle and Ukhwi, the level of resident participation and involvement was examined. Only 65% of respondents indicated awareness of the existing tourism projects in Ngwatle and Ukhwi. Moreover, a low level of participation among the locals was noted, especially in Ngwatle where a lack of significant benefits was evident. However, more than 60% of the locals in Ukhwi were active participants in the tourism industry. This discrepancy might be explained by the concentration of tourism-related infrastructure in Ukhwi. The lower literacy rate among residents in Ngwatle might also limit their opportunities to participate in CBO projects. The inequitable distribution of tourism benefits has led to discontent within the Ngwatle settlement. In addition, cultural issues were likely to be of concern. Based on focus group discussions, it was revealed that the San/Basarwa, who are more highly concentrated in Ngwatle, felt marginalized by the Community Trust, which was seen as belonging to the wealthier ethnic Bangologa and Bakgalagadi residents. Furthermore, focus group discussions suggested that there was insufficient consultation at the grassroots level even to create a Community Trust.

Management of community trust

CBOs such as the NKXT are created, owned and operated by local communities. There is an inherent expectation that local people would obtain a sense of ownership and be satisfied with the management of their CBO, yet results revealed surprisingly low levels of satisfaction with the Trust. Ninety-five per cent of respondents in Ngwatle and 74% of respondents in Ukhwi reported dissatisfaction. This was closely aligned to perceptions of an unequal distribution of benefits from the CBO. Ninety per cent of respondents in Ngwatle and 62% in Ukhwi felt that benefits were not equitably distributed. Despite positive expectations during the inception phase of the Trust, the communities of Ukhwi, and particularly Ngwatle, appear to have lost confidence in their CBO. The failure of the Trust was due to poor management, lack of funding, general lack of participation by the community residents, breach of the initial joint venture contract with the safari operator and lack of technical guidance. All these factors have manifested in dissonance and ethnic mistrust between the communities. Overall, the trust did not generate sufficient profits to sustain existing village-based tourism activities, which resulted in low levels of tourism activity and gradual deterioration of community projects.

The apparent community discord that has arisen due to the perceived elitism of the Trust management has been detrimental to the communities. The failure to meet community expectations might also be explained in part by the withdrawal of an international NGO that abandoned operations prior to the expiration of its two-year contract (NKXT, 1999; van der Jagt *et al.*, 2000). Community participation has not increased since the withdrawal of the NGO. The divisiveness that has followed and the community's low capacity to resolve conflict and negotiate change are threats to the sustainability of community village-based tourism in the KD1 area.

Conclusions

This chapter has highlighted several problems and opportunities related to the development of village-based tourism. The case study communities of Ngwatle and Ukhwi are dependent on government welfare programmes including destitution benefits, aged pensions and orphanage rations. Employment opportunities are largely realized through the Local Government Drought Relief projects (Chanda & Totolo, 2003). Despite the aims of CBNRM projects to empower local communities and reduce dependency, the establishment of the NKXT has had little impact on the level of government dependency and/or on the level of reliance on hunting and gathering for subsistence.

Selebatso (2005: 1) claimed that 'the full devolution of natural management responsibilities to local communities from NGOs and governments instills a sense of resource ownership and drives towards goals of sustainable wildlife conservation'. However, this study has shown little participation in decision-making by the local communities in the NKXT. Furthermore, similar patterns were reported in other districts with projects such as the Khawa Community Development Trust (Arntzen *et al.*, 2003; van der Jagt *et al.*, 2000).

Another key issue in Botswana is that failure of projects following the withdrawal of donor organizations has been relatively common (Arntzen *et al.*, 2003). It is apparent that community projects in the country require longer periods of support in order to become sustainable. Additionally, poor planning and a lack of knowledge (e.g. tourism expertise and marketing skills) and coordination are common problems among CBOs in Botswana, where conflicts have arisen as a result of mismanagement and misappropriation of funds (BOPA, 2007). The CBNRM concept is not without its successes, but there is a need for communities to identify and refine ways of channelling proceeds from CBOs to the people (Arntzen *et al.*, 2003; Garrod, 2003; Schuster & Thakadu, 2007; Telfer, 2002; Timothy, 2003).

Overall, there is still a high dependence on safari hunting activities with little development of non-consumptive tourism such as photographic safaris and cultural tourism activities. Nevertheless, there is considerable potential for economic development via village tourism in the Kalahari region. To create employment opportunities, there is a need to promote the craft industry through the establishment of craft centres or cooperatives. The craft that is produced is high-quality traditional bead-work, bone or wood carving and leatherwork. Currently, low levels of craft production are evident, limiting tourism revenue generation. With assistance from craft centres, local communities will need to be empowered to market locally developed products that can be packaged for both domestic and international consumption. There is a major need for tourism- and business-related education for residents to help them make informed decisions with regard to tourism development and future directions. Such training should be ongoing to encourage continual improvement in both product and service elements of tourism provision. Other efforts needed to enhance potential include raising awareness of tourism and its benefits, improving tourism-related infrastructure, developing local skills for meaningful participation and marketing of the attractions in the community and the Kalahari region (see Mbaiwa *et al.*, Chapter 12 in this book).

Given the wealth of tourist attractions in the form of wildlife, beautiful salt pans, authentic village architecture and traditional cultures, a strategic marketing focus by the Botswana Tourism Board in conjunction with the community could stimulate demand. Also, the term 'Kalahari' is an identifiable brand globally, which needs to be leveraged to promote the region for tourism. While the safari product is lucrative and should be utilized, it is unlikely to be sustainable. It is essential that non-consumptive, natural and cultural tourism be developed and promoted to diversify the tourism products in the region (see Atlhopneng & Mulale, Chapter 8 in this book).

Finally, although the benefits of community- and village-based tourism initiatives have not been fully realized in the KD1 area of the Matsheng villages, the potential remains strong. The prime challenge is to empower communities through tourism and business education, and to facilitate cooperation and trust between different ethnic groups in the communities. Extended support from governments and NGOs is required, but it should be geared towards self-dependency, capacity building and sustainability.

References

Arntzen, J.W., Molokomme, D.L., Terry, E.M., Moleele, N., Tshosa, O. and Mozambani, D. (2003) *Main Findings of the Review of CBNRM in Botswana* (Occasional Paper No. 24). Gaborone: CBNRM Support Programme.

Ashley, C. and Jones, B. (2001) Joint ventures between communities and tourism ventures: Experience in Southern Africa. *International Journal of Tourism Research* 3 (3), 403–423.

Barnes, J.I. (2001) Economic returns and allocation of resources in the wildlife sector of Botswana. *South African Journal of Wildlife Research* 31 (3 and 4), 141–153.

Botswana Press Agency (BOPA) (2007) Community based organizations must be empowered. *Botswana Daily News*, Botswana.

Botswana Review (2005) *Botswana Review* (25th edn). Gaborone, Botswana: Botswana Export Development & Investment Authority (BIDPA).

Botswana Tourism Development Program (BTDP) (1998) *Visitor Survey*. Ministry of Commerce and Industry, Department of Tourism, Gaborone, Botswana.

Botswana Tourism Development Program (BTDP) (1999) *Tourism Economic Impact Assessment*. Ministry of Commerce and Industry, Gaborone, Botswana.

Burns, P.M. and Barrie, S. (2005) Race, space and 'Our own piece of Africa': Doing good in Luphisi village? *Journal of Sustainable Tourism* 13 (5), 468–485.

Campbell, B., Byron, N., Hobane, P., Madzudzo, E., Matose, F. and Wily, L. (1999) Moving to local control of woodlands resources – can CAMPFIRE go beyond the mega-fauna? *Society & Natural Resources* 12 (5), 501–509.

Central Statistics Office (2007) *Quarterly Reports*. Harare: Central Statistics Office, Government of Zimbabwe.

Chanda, R. and Magole, L. (2001) Rangelands in the context of subsistence livelihoods: The case of the Matsheng Area, Kgalagadi North, Botswana. *Global Change and Community Rangelands in Southern Africa* (Working Paper No. 3). Gaborone, Botswana.

Chanda, R. and Totolo, O. (2003) Environmental change and sustainability issues in Kalahari Region. *Journal of Arid Environments* 54 (2), 257–259.

Chanda, R., Totolo, O., Moleele, N., Setshogo, M. and Mosweu, S. (2003) Prospects for subsistence livelihood and environmental sustainability along Kalahari Transect: The case study of Matsheng in Botswana's Kalahari rangelands. *Journal of Arid Environments* 54 (2), 425–445.

Conyers, D. (1990) Centralisation and development planning: A comparative perspective. In P. de Valk and K.H. Wekwete (eds) *Decentralising for Participatory Planning* (pp. 15–34). Newcastle Upon Tyne: Athenaem Press Ltd.

Cottrell, S.P. (2001) A Dutch international development approach: Sustainable tourism development. *Parks and Recreation* 36 (9), 86–92.

Dei, L.A. (2000) Community Participation in tourism in Africa. In P.U.C. Dieke (ed.) *The Political Economy of Tourism in Africa* (pp. 285–298). New York: Cognizant.

Department of Wildlife and National Parks (DWNP) (1999) *Joint Venture Guidelines*. Gaborone, Botswana.

Dieke, P.C. (ed.) (2000) *The Political Economy of Tourism in Africa*. New York: Cognizant.

Ditlhong, M. (1997) *Poverty Assessment and Poverty Alleviation in Botswana* (Working Paper No. 12). Gaborone, Botswana: Botswana Institute for Development Policy Analysis (BIDPA).

Dyer, P., Aberdeen, L. and Schuler, S. (2003) Tourism impacts on an Australian indigenous community: A Djabgay case study. *Tourism Management* 24 (1), 83–95.

Frost, W. (2003) The financial viability of Heritage Tourism attractions: Three cases from rural Australia. *Tourism Review International* 7 (1), 13–22.

Garrod, B. (2003) Local participation in the planning and management of ecotourism: A revised model approach. *Journal of Ecotourism* 2 (1), 33–53.

Granlund, L. (2001) The abundance of game in relation to villages, livestock and pans – a study in southwestern Kalahari, Botswana. MSc thesis, Field Study 73, Uppsala University.

Government of Botswana (GOB) (1986) *Wildlife Conservation Policy* (Government Paper No. 1). Gaborone, Botswana: Government Printer.

Government of Botswana GOB (1990) *Tourism Policy* (Government Paper No. 2 of 1990). Gaborone, Botswana: Government Printer.

Government of Botswana (GOB) (1992) *Wildlife Conservation and National Parks Act* (Paper No. 28 of 1992). Gaborone, Botswana: Government Printer.

Government of Botswana (GOB) (1998) *Review of Remote Area Development Programme*. Gaborone, Botswana: Government Printer.

Government of Botswana (GOB) (2001) *Botswana National Atlas*. Botswana: Department of Surveys and Mapping.

Government of Botswana (GOB) (2002) *The National Development Plan Nine (NDP 9); NDP 9 for 2003/2004 to 2008/2009*. Gaborone, Botswana: Ministry of Finance and Development Planning.

Government of Botswana (2005) *Draft Community Based Natural Resources Management (CBNRM) Policy* (Government Paper). Gaborone, Botswana: Minister of Environment, Wildlife and Tourism.

Government of Zimbabwe (1992) *Parks and Wildlife Act*. Zimbabwe: Government Printer.

Gumbo, D.J. (1992) Traditional management of indigenous woodlands. A paper presented at the Wood biomass and use patterns in communal areas of Zimbabwe, Harare, 26–29 May.

Gujardhur, T. (2001) *Joint Venture Options for Communities and Safari Operations in Botswana* (Occasional Paper No. 6). Botswana: SVN/IUCN, CBNRM Support Programme.

Hasler, R. (1996) Brokering the community basis of tourism potential in the Hwange Communal Lands. CASS Occasional Paper Series, NCRM 1995.

Hasler, R. (1998) The political and socio-economic dynamics of natural resource management: The communal areas management programme for indigenous resources in Chapoto Ward 1989–1990. CASS Working Paper, NRM Series, CPN 12, 91.

Hoole, A. (2001) *Integrated Resources Assessment and Environmental Management Guidelines for the Zambezi River above the Victoria Falls*. Victoria Falls: Government of Zimbabwe Department of Physical Planning.

Hitchcock, R.K. (1997) Cultural, economic and environmental impacts of tourism among Kalahari Bushmen. In E. Chambers (ed.) *Tourism and Culture: Applied Perspective*. Albany: State University of New York Press.

Holloway, S.L. (2007) Burning issues: Whiteness, rurality and the politics of difference. *Geoforum* 38, 7–20.

Hong, W. (1998) *Rural Tourism: A case study of Regional Planning in Taiwan. The International Information center for the farmers in the Asia Pacific Region*. On WWW at http://www.agnet.org/library/article/eb456. Accessed on July 2007.

Intrepid Travel and Victoria University (2002) *Small Tour Group Impacts on Developing Communities*. On WWW at http://www.intrepidtravel.com/about/allabout/rt/research.php.Accessed on June 2007.

Johnson, P. (1996) *Tourism Development Plan for Kgalagadi District*. Botswana: Phillip Johnson Associates.

Joppe, M.C. (1996) Sustainable community tourism development revisited. *Tourism Management* 17 (7), 475–479.

Madzudzo, E. (1998) *Community-based Natural Resource Management in Zimbabwe: Opportunities and Constraints* (CASS Working Paper, NRM Series, CPN 101/98). Harare: CASS UZ.

Makumbe, J.M. (1996) *Participatory Development*. Harare: University of Zimbabwe.

Makumbe, J.M. (1998) *Democracy and Development in Zimbabwe: Constraints of Decentralization*. Harare: Southern Africa Regional Institute for Policy Studies.

Mberengwa, I. (2000) *Small-Scale Indigenous Rural Communities and Community-based Natural Resource Management Programmes in Marginal Areas of Zimbabwe. A Study in Sustainability*. Harare: Centre for Applied Social Studies (CASS).

Ministry of Local Government (MLG) (1997–2003) *Kgalagadi District Development Plan 4* (DDP 4). Gaborone, Botswana.

Ministry of Local Government (MLG) (1997–2003) *Kgalagadi District Development Plan 5* (DDP 5). Gaborone, Botswana.

Ministry of Local Government (MLG) (2003–2009) *Kgalagadi District Development Plan 6*: 2003/4. Gaborone, Botswana.

Motavalli, J. (2002) Taking the natural path. *The Environmental Magazine* 13 (4), 26–37.

Moon, O. (1989) *From Paddy Fields to Ski Slope: The Revitalization of Tradition in Japanese Village Life*. Manchester: University Press.

Murphy, P. (1985) *Tourism a Community Approach*. London: Methuen.

Mutizwa-Mangiza, N.D. (1990) *Decentralization in Zimbabwe: Problems of Planning at the District Level* (RUP Occasional Paper No. 16). Harare: University of Zimbabwe.

Nepal, S. (2005) Tourism and remote mountain settlements: Spatial and temporal development of tourist infrastructure in the Mount Everest Region, Nepal. *Tourism Geographies* 7 (2), 205–227.

Ngwarai, K. (2000) Development of citizen hunting in protected areas. A paper presented at the Sustainable Tourism Development in Protected Areas Systems, held in Kwazulu Natal, South Africa, 6–8 March.

Nqwaa Khobee Xeya Trust (NKXT) (1999) Land Use Management Plan. Kgalagadi, Botswana.

Nyaupane, G., Morais, D. and Dowler, L. (2006) The role of community involvement and number/type of visitors on tourism impacts: A controlled comparison of Annapurna, Nepal and Northwest Yunnan, China. *Tourism Management* 27 (6), 1373–1385.

Phuthego, T.C. and Chanda, R. (2003) Traditional ecological knowledge and community based natural resources management. Lessons from a Botswana Wildlife management area. *Journal of Applied Geography* 24 (10), 57–76.

Rattanasuwongchai, N. (1998) *Rural Tourism – The Impact on Rural Communities II*. Thailand. On WWW at http://www.fftc.agnet.org/library/data/eb458b/eb458b.pdf. Accessed on September 2007.

Roberts, L. and Hall, D. (eds) (2001) *Rural Tourism and Recreation: Principles to Practice*. London: Routledge.

Rondinelli, D.A. (1993) *Development Projects as Policy Experiments*. London: Routledge.

Roodt, V. (2004) *The Shell Tourist Travel Guide of Botswana* (3rd edn). Gaborone: Shell Oil Botswana.

Selebatso, M. (2005) The Role of CBNRM in Wildlife Conservation in Southern Africa. *Tropical ecology and Management*. Department of Ecology and Natural resources management, Norwegian University-on-line Review. On WWW at http://www.umb.no/ina/studier/sopgaver/2005_selebatso.pdf. Accessed on June 2006.

Schuster, B. and Thakadu, O.T. (2007) *Natural Resources Management and People* (Occasional Paper No. 5). Botswana: CBNRM Support Programme.

Sofield, T. (1991) Sustainable ethnic tourism in the South Pacific: Some principles. *The Journal of Tourism Studies* 2 (1), 156–172.

Taylor, M. (2007) *CBNRM for Whose Benefit? A Case Study of Subsistence Hunting on the Boundaries of Botswana's Northern Protected Areas* (Occasional Paper No. 5). Botswana: CBNRM Support Programme.

Telfer, D.J. (2002) The evolution of tourism and development theory. In R. Sharpley and D. Telfer (eds) *Tourism and Development Concepts and issues* (pp. 35–78). Clevedon: Channel View Publications.

Thai, M.T. (2002) *Does Tourism Help Local People: A Vietnam Focus*. On WWW at http://www2.Hawaii.edu/~mthai/papers/tourism. Accessed on September 2007.

Timothy, D.J. (2002) Tourism and community development issues. In R. Sharpley and D. Telfer (eds) *Tourism and Development Concepts and Issues* (pp. 149–164). Clevedon: Channel View Publications.

Van der Jagt, C.J., Gujadhur, T. and Van Bussel, F. (2000) *Community Benefits through Community Based Natural Resources Management in Botswana* (Occasional Paper No. 2). Gaborone, Botswana: CBNRM Support Programme.

Virtanen, P. (2003) Local management of global values: Community-based wildlife management in Zimbabwe and Zambia. *Society and Natural Resources* 16, 179–190.

Wekwete, K. and Mlalazi, A. (1990) Provincial/regional planning in Zimbabwe: Problems and prospects. In P. de Valk and K.H. Wekwete (eds), *Decentralising for Participatory Planning* (pp. 73–85). Newcastle Upon Tyne: Athenaem Press Ltd.

World Travel and Tourism Council (WTTC) (2007) *Botswana: The Impact of Travel and Tourism on Jobs and the Economy*. London: WTTC.

WTO (2005) At http://www.world-tourism.org/newsroom/Releases/2005/july/prwto.htm. Accessed on July 2005.

Ying, T. and Zhou, Y. (2007) Community, governments and external capitals in China's rural cultural tourism: A comparative study of two adjacent villages. *Tourism Management* 28 (1), 96–107.

Zeppel, H. (2002) Cultural tourism at the Cowichan Native Village, British Columbia. *Journal of Travel Research* 41 (1), 92–100.

Chapter 12

The Socio-economic Impacts of Tourism in the Okavango Delta, Botswana

JOSEPH E. MBAIWA and MICHAEL B.K. DARKOH

Introduction

The tourism industry has been one of the global economic success stories in the last 40 years (Coccossis & Parpaires, 1995). In Botswana, tourism was almost non-existent at independence from British rule in 1966. However, by 2007, it had grown to be the second largest economic sector contributing about 9.5% to the country's gross domestic product (GDP) and just under 16% of the non-mining GDP (World Travel & Tourism Council, 2007). Most of Botswana's holiday tourists visit the Okavango Delta, an inland wetland and rich wildlife habitat, located in northwestern Botswana (Figure 12.1).

Tourism development in the Okavango Delta has stimulated the development of a variety of allied infrastructure and facilities, such as hotels, lodges and camps, airport and airstrips. Through its backward linkages, wholesale and retail businesses have also been established, especially in Maun (the main tourist centre in the Okavango), to offer various goods and services to the tourist industry. Tarred roads and other communication facilities have been developed in the region partly to facilitate tourism development. These tourism facilities and services have led to a booming tourism and related economy built around what is perceived internationally as a 'new' and 'exotic' destination. In addition to these positive socio-economic impacts, there are negative aspects of tourism development in the Okavango Delta which are also highlighted in this chapter.

Sustainable Tourism Development

Sustainable tourism development, ecotourism, alternative tourism and green tourism are some of the frameworks that have emerged in the last

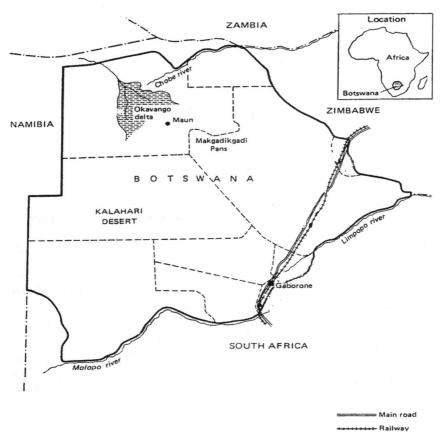

Figure 12.1   Map of Botswana showing the Okavango Delta
Source: Mbaiwa (2005)

two or three decades aiming at promoting a tourism industry that satisfies the needs of the present generations without compromising the ability of future generations from benefiting from the same resources (see Saarinen, Chapter 5 in this book). Tourism development has four forces of social change that drive sustainability (Liu, 2003; Prosser, 1994). These are dissatisfaction with existing products, growing environmental awareness and cultural sensitivity, realisation by destination regions of the precious resources they possess and their vulnerability and the changing attitudes of developers and tour operators. Coccossis (1996) argues that sustainable tourism is a tourism approach that adopts three principles of sustainable development, namely, economic efficiency, environmental conservation and social equity (Figure 12.2) in its development.

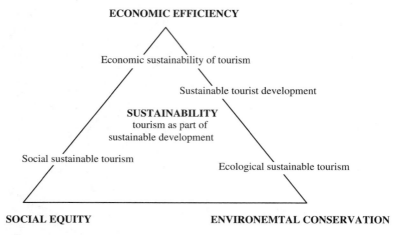

Figure 12.2 Interpretation of sustainable tourism
Source: Modified from Coccossis (1996)

Tosun (2001: 290–291) argues that sustainable tourism development should meet the following conditions:

- Contribute to the satisfaction of basic and felt needs of those hitherto excluded in local tourist destinations.
- Reduce inequality and absolute poverty in local tourist destinations.
- Contribute to the emergence of necessary conditions in tourist destinations which will lead local people to gain self-esteem and to feel free from the evils of want, ignorance and squalor.
- Help host communities to be free or emancipated from alienating material conditions of life and from social servitude to nature, ignorance, other people, misery, institution and dogmatic beliefs.
- Accelerate not only national economic growth, but also regional and local economic growth which must be shared fairly across the social spectrum.
- Achieve the above principles indefinitely without compromising the ability of future generations to meet their own need.

Tosun (2001) says that failure of tourism development to satisfy these prime requirements or objectives renders the industry unsustainable. Therefore, any analysis of the socio-economic impacts of tourism development in wilderness areas such as the Okavango Delta should be based on these principles of sustainable tourism development.

Positive Socio-economic Impacts of Tourism

As noted above, one of Tosun's (2001) conditions for sustainable development is the acceleration not only of national economic growth, but also regional and local economic growth which must be shared fairly across the social spectrum. In the Okavango Delta, infrastructure development in the form of the network of roads, Maun Airport and airstrips, hotels, lodges and safari camps can be described as promoting the national, regional and local economic growth where all stakeholders living in these areas benefit from these economic and social facilities and services. Infrastructure development in the Okavango is critical in meeting the goals of sustainable tourism as all movement of people and commodities in the tourism industry depends on the availability and efficiency of the transportation network and other service facilities. Infrastructure development and service facilities in the Okavango are elaborated as follows:

The road network in northern Botswana

Good roads especially those with tarmac were virtually non-existent in the Okavango region before the 1990s. However, the situation gradually changed especially after the government realised that tourism in northwestern Botswana had the potential of positively contributing to and diversifying the country's economy and adopted the Tourism Policy of 1990. As a result, the construction of tarred roads to facilitate the tourism industry in the Okavango became one of the major infrastructure development priorities for the government. Table 12.1 shows some of the road networks in the Okavango that have since been tarred to improve the road linkage.

The tarred road network in the Okavango Delta especially that from Shorobe through Maun to Mohembo has facilitated the easy movement of mobile and self-drive tourists into the Delta as well as promoting quick delivery of tourist supplies to camps and lodges within the Delta. The road network in the Okavango Delta is an important factor in tourism development as accessibility and movement of goods and services in the

Table 12.1   Road network linking Okavango region

Road link	Kilometres	Year of completion
Francistown–Kasane	505	Late 1980s
Nata–Maun	304	1992
Maun–Sehitwa	93	1993
Gumare–Sepupa	73	1992
Sepupa–Mohembo	80	1994
Gumare–Tsau	117	1990
Maun–Shorobe	37	1991
Ghanzi–Sehitwa	200	2000

Source: Darkoh and Mbaiwa (2006)

Okavango have considerably improved. As a result, the needs of tourists who visit the Okavango, safari companies which operate tourism businesses, government and local people who derive income from tourism through taxes or employment opportunities are benefiting from tourism development. It can therefore be argued that some distribution of benefits to stakeholders as espoused by sustainable tourism principles is somehow occurring in the Okavango Delta in as far as the development and use of the road network is concerned.

Maun International Airport

Air transport has become an important means for travelling and achieving the goals of sustainable tourism in as far as promoting the national, regional and local goals is concerned. As such, tourism facilities like Maun Airport play a major role in the facilitation of tourism development and regional and local development in the Okavango Delta. Maun Airport provides an air link between Maun and Johannesburg, Windhoek, Harare, Victoria Falls and Gaborone. It also provides a link with about 50 airstrips located within the Okavango Delta. The Maun Airport has in the recent past been described as one of the busiest airports in Botswana and Africa especially during tourist peak seasons. According to Ngami Times Newspaper (2001: 1), '. . . Maun Airport is . . . regarded as the second busiest international in Africa in terms of aircraft movements after the combined Johannesburg area (South Africa) airports of Johannesburg International, Lanseria, Rand and Grand Central'.

In 2000, Maun Airport had an average of 256 aircrafts landing and taking off each day during the tourist peak seasons of April to October. In the non-tourist peak seasons of November to March, the figure is reduced to an average of 152 aircrafts landing and taking off each day. A total of 40,246 aircraft movement was recorded at Maun Airport in 2000 (Mbaiwa, 2003). Over 86.1% of the total movements were domestic movements and were made by small engine aircrafts that fly into the delta either transporting tourists or carrying supplies or returning to Maun from the Delta for parking. At present, the government intends to expand Maun Airport in order to allow it to have direct flights with Europe, North America, Asia and Australia where most of the tourists that visit the Okavango originate. This therefore shows the role that air transport plays in promoting economic development in host regions, which is one of the major goals of sustainable tourism development.

Hotels and safari camps in Okavango Delta

The growth of tourism in the Okavango Delta is directly associated with the proliferation of hotels, lodges and safari camps in the Okavango

Delta. There are about 60 photographic lodges and camps in the Okavango region with about 554 rooms and 1018 bed spaces. It is in these accommodation facilities that most people in the Okavango Delta are employed. Accommodation facilities are important to the socio-economic life of the Okavango and Botswana because of the significant amount of revenue they generate and the number of citizens they employ. While accommodation facilities in the Okavango swamps have been accused for failing to cater for citizens due to high prices paid in these facilities, hotels and lodges in Maun are serving the host population in terms of restaurant services, accommodation and conference facilities. In addition, these facilities provide employment opportunities for host populations; hence the distribution of benefits from these services is an important aspect of the national, regional and local economic growth. A phenomenon which any advocate of sustainable development principles would strive to achieve in host countries.

Rural Development

Local participation in tourism development

Rural development is one of the main impacts of tourism development in the Okavango Delta. Sustainable tourism development advocates for the fairness in the distribution of costs, benefits, decision-making and management by all user groups, this in theory has the potential to eradicate poverty in rural communities (UNCED, 1992). As such, local participation in the costs, benefits, decision-making and management of tourism development in the Okavango Delta is critical in achieving rural development and sustainable tourism development. This goal is achieved through the Community-based Natural Resource Management (CBNRM) programme in the Okavango Delta. The CBNRM programme has been made possible through the adoption of several government policies and strategies, notably, the Wildlife Conservation Policy of 1986, the Tourism Policy of 1990, the National Conservation Strategy of 1990, the Tourism Act of 1992 and the Wildlife Conservation and National Parks Act of 1992 (see also Atlhopneng & Mulale, Chapter 8 in this book). These strategies laid the adoption of the CBNRM programme not only in the Okavango Delta but also in Botswana as a whole. Each of these documents advocates for increased opportunities for local communities to benefit from wildlife and other natural resources through tourism development.

As a result, local communities in the Okavango Delta have formed institutions to ensure their participation in natural resource management and tourism development. These institutions are known as community-based organisations (CBOs) or Trusts. Trusts or CBOs are a legal requirement by the Botswana government before a tourism licence is issued and land and its natural resources are allocated to a community. A Trust can be composed of one or more villages and this depends on geographical location and availability of land with wildlife resources. Trusts therefore provide leadership in the use of land, wildlife and veld resources for tourism purposes. In 2001, there were 12 trusts in the Okavango Delta, that is, about 27% of all trusts in Botswana (National CBNRM Forum, 2001).

The Wildlife Conservation Policy of 1986 led to the demarcation of land in the Okavango Delta into small land units known as controlled hunting areas (CHAs). On average, a CHA covers about 1000 km^2. The government allocates a wildlife quota to a community in a CHA after an annual assessment of wildlife statistics in the area is carried out. It is in these small land units that local communities control and have access not only to land and wildlife resources for tourism purposes, but also to veld products such as reeds and grass that are commonly used for thatching in some of the lodges in the Okavango. For example, the Khwai community cuts thatching grass and then sells this to operators for thatching lodges in the delta. Local communities can both sub-lease their CHA and sell their wildlife quota to safari operators for photography and safari hunting or operate the business for themselves. Access to land and its natural resources for tourism purposes and the formation of institutions (CBOs) can therefore be described as socio-cultural benefits and a form of empowerment to local communities in tourism development in the Okavango Delta. These benefits are in line with the sustainable development ideals of social equity and access to resource use by stakeholders, particularly rural people (see WECD, 1987). Local participation in tourism development through the CBNRM programme and benefits to communities derived from it aim at eradicating rural poverty. This therefore in line with Tosun's (2001) principle of sustainable tourism must make local people gain self-esteem and freedom from want, ignorance and squalor.

Creation of employment

Employment of host populations in tourism is a key factor in achieving sustainable tourism development. This is because employment creates conditions that reduce poverty (Tosun, 2001) because it enables people to achieve basic human needs. Employment in the tourism industry in the Okavango Delta is influenced by the degree of linkages the industry has with other sectors of the economy, notably agriculture, manufacturing, the craft industry, the wholesale and retail industry, the transport sector, especially the airline services, communication and the wildlife sector. While a significant number of people are employed in tourism and associated sectors, the industry has not effectively managed to promote the agricultural, craft and manufacturing sectors (Mbaiwa, 2002). This means a sizable number of people who would otherwise be employed in these sectors are not employed. Mbaiwa (2002) found that a total of

923 people were employed in a sample of 30 lodges and camps in the Okavango Delta and 727 people had jobs in other 35 tourism-related businesses in Maun in 2001. A survey conducted by Scott Wilson Consultants at the same time as Mbaiwa's study but on different camps and lodges found that 735 people were employed in a total of 20 safari camps in the Delta (Scott Wilson Consultants, 2001). In these two surveys, therefore, a total of 50 safari camps and lodges in the Okavango Delta employed about 1658 people, which is 16.6% of the formal employment in the tourism sector of Botswana.

The implementation of the community

The CBNRM programme in the late 1990s increased formal employment opportunities in several villages in the Okavango (Table 12.2). For example, at Sankoyo village (with a population of 372), the number of people employed by the community Trust (STMT) and their Joint Venture Partner (JVP) increased from 51 people in 1997 when the Trust started operating to 108 people (111.7%) in 2007. At Mababe (with a population of 300), the number increased from 52 people in 2000 to 66 people (26.9%) in 2007 and at Khwai (with a population of 290), the number increased from 5 people in 2000 to 74 people (225%) in 2007.

Table 12.2   Employment at Sankoyo, Khwai and Mababe, 1997–2007

	Sankoyo			*Khwai*			*Mababe*		
Year	*Trust*	*JVP*	*Total*	*Trust*	*JVP*	*Total*	*Trust*	*JVP*	*Total*
1997	10	41	51	NA	NA	NA	NA	NA	NA
1998	11	51	62	NA	NA	NA	NA	NA	NA
1999	11	51	62	NA	NA	NA	NA	NA	NA
2000	11	51	62	5	NA	5	15	37	52
2001	13	9	22	5	NA	5	15	64	79
2002	35	56	91	8	NA	8	16	64	80
2003	42	56	98	10	NA	10	18	64	82
2004	48	56	104	15	40	55	41	18	59
2005	45	56	101	15	50	65	41	25	66
2006	46	56	102	19	55	74	41	25	66
2007	52	56	108	19	57	74	41	25	66

Source: Mbaiwa (2008)

These increases are significantly high especially considering that these are tiny villages in which a large proportion of their populations comprise mostly elderly people and school-going children. CBNRM is the most important economic activity that provides formal employment opportunities in these three villages. Since there is neither industrial development nor manufacturing industries in the Okavango region, tourism has become a major employing sector in the area, hence contributing to the goals of sustainable tourism through poverty eradication.

Revenue generated through community-based tourism

Tourism in the Okavango Delta has successfully generated considerable income for local communities through community-based tourism projects. Revenue generation by local communities through the CBNRM programme contributes to the satisfaction of basic and felt needs of host communities as well as rural economic development which are principles of sustainable tourism as described by Tosun (2001). In addition, the availability of revenue in rural areas is important in poverty eradication in these areas. Tourism revenue that accrues to communities is largely from the following activities: sub-leasing of the hunting area; sales of wildlife quota (i.e. wildlife quota fees for game animals hunted) and meat; tourism enterprises, for example, lodges and campsites and camping fees and vehicle hires. Income from tourism development accrues to individuals, households and the community at large when it is finally distributed. Table 12.3 shows the financial benefits that respectively accrued to Sankoyo, Khwai and Mababe from the time the projects started operating to 2007.

Table 12.3 shows that land rentals and quotas in each village have increased over time. For example, at Sankoyo, land rentals increased from P285,750 in 1997 to P483,250 in 2006 and game quota fees increased from P60,928 in 1998 to P1,198,700 in 2006. These findings indicate that between 2004 and 2006, game quota fees are the largest source of revenue for the STMT, accounting for almost half of the revenue generated by the Trust. The revenue that has been accumulating to STMT is substantial and has

Table 12.3   Revenue generated by Sankoyo, Khwai and Mababe, 1997–2007

Year	Land rental	Quota	Others[a]	Total
Sankoyo Tswaragano Management Trust (STMT)				
1997	285,750	0	12,665	298,415
1998	116,666	60,928	38,826	216,420

(Continued)

Table 12.3 *Continued*

Year	Land rental	Quota	Others[a]	Total
1999	151,667	33,470	76,151	261,288
2000	166,833	49,090	148,940	215,923
2001	57,047	55,600	114,801	227,448
2002	492,000	872,550	131,844	1,496,394
2003	466,509	965,772	370,352	1,802,633
2004	562,655	1,096,377	75,634	1,734,666
2005	455,000	1,060,400	612,012	2,127,412
2006	483,250	1,198,700	639,116	2,321,066
2007	613,360	1,272,600	621,537	2,507,497
Khwai Development Trust (KDT)				
2000	1,057,247	0	72,536	1,129,783
2001	585,220	0	248,305	833,525
2002	1,211,533	0	36,738	1,214,567
2003	348,778	0	97,480	446,258
2004	110,000	857,085	283,482	1,250,567
2005	115,500	1,043,707	405,247	1,564,454
2006	121,275	1,248,500	1,248,500	1,691,723
2007	127,339	1,217,187	1,082,146	2,426,667
Mababe Zokotsama Trust (MZT)				
2000	60,000	550,000	77,000	687,000
2001	69,000	632,500	127,233	828,733
2002	79,350	702,606	85,961	867,917
2003	91,205	807,996	98,854	1,121,427
2004	104,940	929,196	149,159	1,183,295
2005	120,681	1,068,575	130,739	1,319,995
2006	120,000	1,202,183	13,500	1,335,683
2007	130,000	1,202,183	29,950	1,362,133

[a]Camp rental fees for Letswee camp, camping fees & community development fund (for Khwai Development Trust); Community development only (for Mababe Zokotsama Trust); Revenue from net profit from Kazikini & Santawani, meat sales, camping, and; vehicle hire donations and community development (for Sankoyo Tswaragano Management Trust).
Source: Mbaiwa (2008)

been on the increase between 1997 and 2007 (Table 12.3). Arntzen *et al.* (2007: 44) argue that 'if one recalls the commercial revenues of the entire Botswana CBNRM program, it becomes clear that STMT is one of the big earners and employers, accounting for well over 10% of the total revenues and employment creation'. This shows that the STMT can be described to be one of the community tourism enterprises that have been achieving some of their major objectives of rural development, particularly the improvement of livelihoods. Part of the money which communities generate from tourism development is reinvested in community development projects such as recreational facilities (e.g. sports ground and community halls), vehicles for transport, lodges, camp sites, small general dealers, bars and bottle stores, as well as payment of salaries of employees in Trusts. This therefore indicates the positive socio-cultural impact of tourism in the Okavango Delta.

Financial Contribution

According to Tosun (2001), sustainable tourism should contribute to the national economic growth of host countries. Revenue generation from Moremi Game Reserve (Table 12.4) is important to the national economic growth since much of the funds generated from it are deposited into the national treasury account to benefit the entire nation of Botswana. Moremi Game Reserve is located in the heart of the Okavango Delta. In recent decades, the Reserve has become one of the most important tourism destinations in Botswana. It offers photographic tourism activities, which

Table 12.4   Number of tourists and revenue collected at Moremi Game Reserve, 1998–2007

Year	Number of tourists	Revenue generated (Pula[a])
1998	49,556	4,301,275
1999	46,707	4,402,121
2000	30,835	5,698,198
2001	31,076	6,198,232
2002	39,734	8,088,936
2003	39,158	7,335,198
2004	38,422	6,822,836
2005	41,513	7,868,711
2006	45,269	6,911,124

[a]5.00 BWP = 1.0 US$ (by mid-2003).
Source: DWNP (2007)

largely depend on the abundance of wildlife in the area. These activities include game viewing, camping, boating and walking trails. In 1971, only 4500 tourists visited Moremi Game Reserve (DWNP, 1991). This number exponentially increased to about 45,000 tourists in 2006 (see Table 12.1). The amount of revenue that the government generates from tourism at Moremi Game Reserve has also increased over the years from about R785.00 (US$120.00) in 1966 (DWNP, 1991) to P7,868,711 (US$1,573,742) in 2005 (Table 12.4).

The increase in tourist numbers, revenue, tourism facilities and activities at Moremi Game Reserve indicates that wildlife resources are an important resource in tourism development in the Okavango Delta. As a result of their economic value, wildlife resources in the Okavango Delta and Botswana as a whole are centralised and protected by government from illegal harvesting and poaching (i.e. for both commercial or subsistence purposes). The protection of wildlife resources and other natural resources in Moremi Game Reserve meets the ecological principles of sustainable tourism since these natural resources are protected from over-harvesting.

Cultural Rejuvenation

Tourism in the Okavango Delta has also promoted the rejuvenation and preservation of some of the following cultural products and practices:

Traditional villages

The formation of Trusts by local communities for tourism purposes has resulted in some of these communities such as Sankoyo, Seronga and Gudigwa establishing tourist camps and lodges which they call traditional villages. The traditional villages provide services such as accommodation in traditional huts, traditional dishes, music and dance, walking trails and animal tracking. In curio shops located in traditional villages, they sell traditionally made souvenirs such as baskets, wood engraved products and beads. Traditional villages do not only promote and preserve local culture, but also provide employment opportunities for local people. As part of tourism development, traditional villages play a significant role in that they promote and preserve local culture that would otherwise be lost. Traditional villages therefore play a significant role in reviving and preserving the local culture in the Okavango Delta.

Mekoro (dug-in canoes) safaris

The wooden dug-out canoes (*mekoro*) have been used by traditional societies in the Okavango Delta for thousands of years as a mode of transportation. Although *mekoro* are still being used in the Okavango, the introduction of tourism in the region has added another fillip to their use which is *mekoro* safaris. Some of the safari companies such as those running Audi Camp, Gunn's Camp, Crocodile Camp and Oddballs provide *mekoro* safaris as one of the major tourist services. In addition, some 75 canoe drivers (polers) at Seronga took advantage of the popularity of *mekoro* safaris and formed the Okavango Poler's Trust (OPT) with the primary objective of offering *mekoro* safaris to tourists (Mbaiwa, 2002). In less than 10 years, the OPT has since reinvested its financial benefits from *mekoro* safaris by establishing a lodge and campsite along the Okavango River at Seronga. This infrastructure benefits the local people in terms of income and employment opportunities.

The use of *mekoro* for tourist activities in the Okavango Delta is one way in which cultural tourism is being promoted. As a result, the history of the people of the Okavango Delta is preserved while at the same time information is provided to visitors on how the different societies in the area were able to move from one point of the delta to the other. This shows that without visitors, local culture and traditions may be lost. Cooper *et al.* (1998) argue that tourism can stimulate interests in, and conserve aspects of the host's cultural heritage. If tourists appreciate the cultural heritage of a destination, then that appreciation can stimulate the host's pride in their heritage and foster local crafts, traditions and customs. This has been found to be the case with *mekoro* safaris in the Okavango Delta.

Basket production

Basket making is carried out in most parts of the Okavango Delta by the different ethnic groups mainly for the tourism market (Mbaiwa, 2003). Baskets were traditionally produced for household and agricultural use not only in the Okavango area but in most parts of Botswana (Groth *et al.*, 1992; Terry, 1999). Because of the influence of tourism, basket production became commercialised in the last 20–30 years (Terry, 1999). Basket making has resulted in income generation for the rural communities in the Okavango Delta. As a result, basket making is one of the cultural artefacts that play an important role in the development of cultural tourism in the Okavango. It also promotes income diversification and improves rural livelihoods in the area.

Negative Socio-economic Impacts of Tourism

The emergence of enclave tourism

The tourism industry that has so far developed in the Okavango Delta is largely owned and controlled by foreign safari companies and investors (Mbaiwa, 2005). Tourism that develops in remote areas and is largely

owned and controlled by outsiders (e.g. expatriates) has in recent literature been referred to as 'enclave tourism' (Britton, 1981; Ceballos-Lascurain,
1996; Mbaiwa, 2005). Ceballos-Lascurain defines enclave tourism as tourism that is concentrated in remote areas in which the types of facilities and
their physical location fail to take into consideration the needs and wishes
of surrounding communities. The goods and services available are beyond
the financial means of the local communities and any foreign currency created may have only a minimal effect upon the economy of the host location. In enclave tourism, facilities are characterised by foreign ownership
and are designed to meet the needs and interests of foreign tourists
(Ceballos-Lascurain, 1996).

   As tourism in the Okavango Delta is enclave in nature, much socio-
economic benefits such as better paying jobs particularly those in management are occupied by expatriates (Mbaiwa, 2005). Glasson *et al.* (1995)
argue that the dominance of the industry by foreign investors and the
non-local investment can reduce control over local resources. Glasson *et al.*
note that the loss of local autonomy is certainly the most negative long-
term socio-cultural effect of tourism in a destination area. A local resident
may also suffer a loss of sense of place, as his/her surrounding is transformed to accommodate the requirements of a foreign-dominated tourism
industry. These claims might be true because interviews with the local
people in the Okavango indicate that there is a general assumption that
the delta has been taken from them by government and given to foreign
tour operators.

   Local communities particularly those involved in CBNRM operate their
businesses in the outskirts of the Okavango Delta. Because of this nature
in tourism development, some citizens view the approach negatively
because they perceive the domination of inner parts of the Delta by non-
citizen companies as 'selling out' of their resources (Mbaiwa, 2005). As
such, there is no equal access to the use of resources and decision-making
between the local people and the tour operators. This is not in line with
the ideals of sustainable tourism which promotes equal access and opportunities to all user groups. Glasson *et al.* (1995) and Ceballos-Lascurain
(1996) suggest that tourism should provide for local participation in decision-
making and the employment of local people to make it sustainable. This
inequality thus provides a challenge for tourism development in
the Okavango Delta to be designed to meet the principles of sustainable
tourism development.

The relocation of traditional settlements

   The Government of Botswana has either relocated or continuously
made suggestions and attempts at relocating some traditional settlements
from the Okavango Delta to promote tourism and wildlife management

activities (Mbaiwa, 2002). These have in the process caused land-use conflicts between rural settlements and wildlife and tourism. The people of Gudigwa and Khwai have been relocated from the inner parts of the Okavango when the area was designated to be Moremi Game Reserve (Mbaiwa & Rantsudu, 2003). The inhabitants of the villages of Jao Flats, Ditshiping and Xaxaba are also being encouraged to relocate from their present sites for tourism and wildlife management purposes.

These kinds of measures by government exacerbate the hostility and conflict that already exists between the local community and the tourist industry. The suggestion to re-locate Khwai village is based on the assumption that wildlife and people cannot co-exist and utilise the same area (Dikobe, 1995). This contradicts the government's strategy of CBNRM that is designed to promote local community involvement in the management of natural resources, thereby ensuring them direct resource benefits from these resources (Government of Botswana, 1986, 1990). These contradictions show the lack of harmonisation in government policies in resource use, hence land-use conflicts among different resource users. Relocation of local communities to give way to tourism development and natural resource management against their will is against the principles of sustainable tourism where development should be carried out with the consent and involvement in the decision-making process of all the interested parties (see also Ramutsindela, Chapter 10 in this book).

Demonstration effect

One of the major impacts of the tourist–host relationship in the Okavango Delta is the demonstration effect. Demonstration effect refers to a situation when the host behaviour is modified to imitate tourists (Ratz, 2002). Tourists visiting the Okavango Delta introduce and display a foreign way of life to the local population. In describing demonstration effect, Jafari (1974) states that local people often tend to imitate the seemingly rich tourists. As a result, a shift in local consumption patterns occur (Wood, 1997) towards western products (Oppermann & Chon, 1997). In the Okavango Delta, foreign tourists and tour operators have influenced the dress code particularly on young people. Young people generally imitate tourists in wearing safari type of clothing such as Khaki pair of trousers/ shorts and shirts, skirts and blouses, which were not common in the area before the introduction of tourism in the Okavango (Mbaiwa, 2004). Some young local females have also been influenced to wear mini skirts and clothes that expose parts of their bodies such as the belly and part of their breasts. Some wear clothes that expose part of their underwear. A manager in one of the main supermarkets in Maun stated that some tourists pass through Maun 'half-naked', meaning that clothes that they wear expose much of their bodies. This manager noted that exposing one's body in public is culturally unacceptable to rural communities especially to the elderly in the Okavango Delta. In this regard, tourism has socio-cultural impacts, ranging from the clothes people wear, to the food people eat and general lifestyles and attitudes which people display.

Prostitution

The problem of prostitution undeniably existed before tourism development in developing countries (Cohen, 1982). However, tourism has undoubtedly contributed to an increase in prostitution in many countries even though it is difficult to determine exactly by how much (Oppermann & Chon, 1997). In the Okavango, prostitution is common in areas which are commonly visited by tourists such as hotels and restaurants. In her survey, Stellenburg (2003) found that prostitutes at Sedia Hotel and Riley's Hotel in Maun are described by managers of these facilities to be a problem. Although Stellenburg found that prostitutes target business people coming from other urban areas of Botswana such as Gaborone, tourists from Europe and North America were the most preferred. This is because tourists from Europe and North America pay more money and at times pay in stronger currencies such as the United States dollar, British pound and European Euro.

Some of the major reasons why women engage in prostitution in the Okavango include the need to increase their income to support their families and pleasure. In the latter case, local black women want to have an experience of having sex with a white person, that is, a European or North American (Stellenburg, 2003). Some of the prostitutes in the Okavango have full-time jobs during the day where they work as secretaries or primary school teachers. Some were found to be University of Botswana students on long winter vacation (Stellenburg, 2003). Although most of the prostitutes were local residents, some were found to be coming from neighbouring countries of Zimbabwe and Zambia (Stellenburg, 2003). The rapid growth of tourism particularly the increase in hotels, lodges and camps in the Okavango and tourists in the Okavango Delta can be assumed to be contributing to prostitution in the area, even attracting prostitutes from neighbouring countries such as Zambia and Zimbabwe which are currently facing economic hardships.

The movement of prostitutes from other countries or from one part of the country to areas where there is a tourism boom is not unusual. Cooper *et al.* (1998) state that, more recently, a major tourism market has grown up around sex tourism in destinations such as Thailand, the Gambia and some of the Central European countries. Commercial sex in these countries may not be officially marketed but tourism in the areas attracts it. Sex tourism or prostitution is not legally allowed in Botswana; however, it continues to be a growing social problem in destination areas such as the

Okavango Delta. The dangers of prostitution are that it is associated with the cultural breakdown and the spread of sexually transmitted diseases such as HIV/AIDS. If a condom is not used, the price is higher. This may not be unique to the Okavango Delta but common in the sex industry in other parts of the world.

Living and working conditions of safari workers

The relationship between workers in the tourism industry on the one hand and their managers and employers on the other in the Okavango Delta can be described as not cordial. This is reflected by accusation and counter-accusations between the two groups (Mbaiwa & Darkoh, 2006). For example, employers and managers accuse workers for laziness, coming to work under the influence of liquor, theft, insubordination and failure to deliver. On the other hand, workers accuse their employers of poor conditions of work (low salaries, long working hours without overtime payments and poor worker accommodation facilities).

Tourism workers complain that they work long hours beyond the eight-hour recommended time period by the Department of Labour. In addition, they are not paid overtime allowances. They are also made to work during public holidays but are not paid overtime as is the case with government workers employed on similar conditions. Workers in the Okavango Delta also complain for being accommodated in small one-roomed and crowded residential tenements which are often used as both kitchen and bedroom (in some cases shared by two people). Some of the rooms were said to be poorly maintained, and saddled with such problems as roof leaks. Other workers noted that they are accommodated in rooms previously built as storerooms which lack the necessary ventilation of a living room or bedroom. Some workers live in tents that are usually not replaced over a long period of time of which some are too old and worn out. This leaves them vulnerable to attacks from wild animals. Bathrooms and toilets are shared and are often located a few metres from bedrooms or tents. Walking to these facilities at night, especially to toilets, also exposes workers to attacks by wild animals. Safari workers also noted that their salaries are low, irrespective of the amount of work they do, the long hours they work and the number of years they have spent at work.

Some of the living and working conditions of safari tourism workers in the Okavango Delta appear to be similar to those of other workers in nature-based tourism destinations in developing countries such as in the Caribbean and Asia or the Middle East. In the Caribbean, Pantin (1998) notes that in St. Lucia, nine out of 10 managers in the hotel and restaurant sectors were expatriates and their average salaries were several times higher than the earnings of the unskilled local labourers. In the Middle East, Seckelmann (2002) notes that the expansion of tourism development in Turkey is dominated by foreign companies. This has resulted in local people providing cheap labour in tourism facilities instead of being the main actors and beneficiaries of the growing tourism business.

All these conditions and experiences therefore indicate that tourism development in many destinations of developing countries fails in many ways to meet the conditions and principles of sustainable tourism. In this regard, the tourism industry becomes an exploiting industry instead of promoting a fair distribution of development as desired by governments and people of host regions.

The expansion of illegal settlements

The growth of tourism in the Okavango Delta is associated with the development of illegal settlements in the area. Thabazimbi Settlement is an example of an illegal and squatter settlement that has developed due to the influence of tourism. The settlement developed around 1983 and has an estimated population of about 150–200 people. The settlement developed as canoe drivers or polers (about 40 in 2001), working in one of the tourism facilities in the Okavango Delta (i.e. for Gunns Camp), were not provided with accommodation (some of these polers freelance with the camp). Residents of Thabazimbi come from various villages around the Okavango Delta. The population of the settlement has been on the increase as more people have been coming to the settlement to find employment in the surrounding camps. Residents of Thabazimbi construct their huts using reeds (a few have used mud), thatching grass and pools. The sanitation in the settlement is very poor as there are no toilets and there is no one to collect the solid waste. The settlement has no water reticulation and schooling services.

The development of illegal settlements in conserved areas is likely to have long-term effects on the socio-economic lives and ecology of the area. For example, the introduction of socio-economic activities such as crop farming and livestock farming is possible in the long run. According to Campbell (1997), most of southern and eastern Botswana was in the 1700s richly endowed with a variety of wildlife. However, wildlife has since disappeared mainly because of the growth of the human population and the introduction of socio-economic activities in the area. A similar situation recently occurred in the Schwelle region where boreholes and cattle farming were introduced in the area. Perkins and Ringrose (1996) note that the various wildlife species which used to exist in this region have since been wiped out by cattle farming. This therefore means if illegal settlements are allowed to spread in the Okavango Delta, there are likely to be serious ecological problems in the area in future. The proliferation of such illegal settlements can compromise the ecological integrity of the Okavango Delta which sustains the tourism industry. Plog (1974) argues that 'tourism contains the seeds of its own destruction, tourism kills tourism, destroying the very environmental attractions which visitors come to a location to experience'. When tourism is dead in the Okavango Delta, this will affect the local and national economy; hence tourism in this regard would have been carried out against the principles of sustainable tourism development.

Conclusion

Sustainable tourism development promotes a tourism industry that achieves national, regional and local economic growth. In the Okavango Delta, tourism development has stimulated the development of a variety of allied infrastructure and facilities, such as hotels, lodges and camps, airport and airstrips, within and around the Okavango Delta. Tarred roads and other communication facilities in and to Okavango have also been developed partly to facilitate tourism development in the region. The other positive impacts of tourism are that the industry provides employment opportunities to local communities and a significant foreign exchange for Botswana. Tourism has become the second largest source of Government revenue in Botswana after the diamond exports. Government revenue from tourism is mostly generated from the Okavango Delta and the Chobe regions. In this regard, tourism development in the Okavango Delta can be noted to be contributing to sustainable tourism development. On the other hand, tourism development in the Okavango Delta has a host of negative socio-cultural and economic impacts that derail the goals of sustainable tourism development. These impacts include enclave tourism, relocation of traditional communities, prostitution and demonstration effect particularly on young people, expansion of illegal settlements, poor living and working conditions of safari tourism workers and relocation of traditional settlements to make room for tourism development.

The negative socio-cultural and economic impacts of tourism development indicate that the expansion of the tourism industry in the Okavango Delta does not give much consideration to issues of sustainable tourism development. Sustainable development is hinged on three main concerns: social equity, economic efficiency and ecological sustainability (Angelson *et al.*, 1994). Social equity advocates for fairness and equal access to resources by all user groups. The aim is to ensure equity in the distribution of costs, benefits, decision-making and management. This assumption is believed to have the potential of eradicating poverty among the poor communities. Contrary to this viewpoint however many local communities in the Okavango have limited access and control over tourism resources. Much of the land and its natural resources such as wildlife that are the main tourist attractions are controlled and owned by either the private tour operators or the government. This has resulted in a lack of a meaningful involvement and participation of the majority of the local people in the tourism business (see also Ramutsidela, Chapter 10 in this book). Major tourism policy decisions are taken without full participation of the local people whose land has been set aside for wildlife and tourism management. This has, as a result, generated conflicts between the local people and the wildlife and tourist industries. Therefore, if tourism in the Okavango Delta is to become sustainable, it should involve local communities in all aspects of its development and promote an alternative livelihood for them.

References

Angelsen A., Fjeldstad, O. and Rashid-Sumaila, U. (1994) *Project Appraisal and Sustainability in Less Developed Countries*. Bergen: Bergen Print Services.

Arntzen, J., Buzwani, B., Setlhogile, T., Kgathi, D.L. and Motsolapheko, M.K. (2007) *Community-Based Resource Management, Rural Livelihoods and Environmental Sustainability*. Gaborone: Centre for Applied Research.

Britton, S. (1981) The spatial organisation of tourism in a neo-colonial economy: A Fiji case study. *Pacific Viewpoint* 21 (2), 144–165.

Campbell, A. (1997) A history of wildlife in Botswana. In *Conservation and Management of Wildlife in Botswana: Strategies for the Twenty-First Century, Proceedings of a National Conference Organised by Kalahari Conservation Society (KCS) and the Department of Wildlife and National Parks* (pp. 7–31). Gaborone: KCS.

Ceballos-Lascurain, H. (1996) *Tourism, Ecotourism and Protected Areas*. Gland: IUCN Publication.

Coccossis, H. and Parpairis, A. (1995) Assessing the interaction between heritage, environment and tourism: Mykonos. In H. Coccossis, and P. Nijkamp (eds) *Sustainable Tourism Development* (pp. 127–140). Hong Kong: Avebury.

Cohen, E. (1982) Thai girls and Farag men: The edge of ambiguity. *Annals of Tourism Research* 9, 403–428.

Cooper, C., Fletcher J., Gilbert, D., Wanhill, S. and Stephen, R. (1998) *Tourism Principles and Practice*. New York: Longman.

Dikobe, L. (1995) People and parks: Do they mix? In Kalahari Conservation Society (KCS) (eds) *The Present Status of Wildlife and its Future in Botswana* (pp. 91–94). Gaborone: Khalahari Conservation Society/The Chobe Wildlife Trust.

DWNP (Department of Wildlife and National Parks) (1991) *Moremi Game Reserve Management Plan*. Gaborone: DWNP.

DWNP (Department of Wildlife and National Parks) (2007) *Annual Progress Reports*. Gaborone: DWNP.

Glasson, J., Godfrey, K. and Goodey, B. (1995) *Towards Visitor Impact Management: Visitor Impacts, Carrying Capacity and Management Responses in Europe's Historic Towns and Cities*. England: Avebury.

Government of Botswana (1986) *Wildlife Conservation Policy* (Government Paper No. 1 of 1986). Gaborone: Government Printer.

Government of Botswana (1990) *The Tourism Policy* (Government Paper No. 2 of 1990). Gaborone: Government Printer.

Groth, A., Jefferies, K. and Woto, T. (1992) *SADCC Rural Industries and Energy Programme, Botswana: Country Report*. Kanye: RIIC.

Jafari, J. (1974) The socio-economic costs of tourism to developing countries. *Annals of Tourism Research* 1, 227–259.

Liu, Z. (2003) Sustainable tourism development: A critique. *Journal of Sustainable Tourism* 11 (6), 459–475.

Mbaiwa, J.E. (2002) *The Socio-Economic and Environmental Impacts of Tourism Development in the Okavango Delta, Botswana: A Baseline Study*. Maun: Harry Oppenheimer Okavango Research Centre, University of Botswana.

Mbaiwa, J.E. (2003) The socio-economic and environmental impacts of tourism in the Okavango Delta, Northwestern Botswana. *Journal of Arid Environments* 54 (2), 447–468.

Mbaiwa, J.E. (2004) The success and sustainability of community-based natural resource management in the Okavango Delta, Botswana. *South African Geographical Journal* 86 (1), 44–53.

Mbaiwa, J.E. (2005) Enclave tourism and its socio-economic impacts in the Okavango Delta, Botswana. *Tourism Management* 26 (2), 157–172.

Mbaiwa, J.E. (2008) Tourism development, rural livelihoods, and conservation in the Okavango Delta. PhD dissertation, Texas A&M University.

Mbaiwa, J.E. and Darkoh, M.B.K. (2006) *Tourism and the Environment in the Okavango Delta, Botswana*. Gaborone: Pula Press.

Mbaiwa, J.E. and Ranstundu, M. (2003) *A Socio-economic Baseline Study of the Bakakhwe Cultural Conservation Trust in the Okavango Delta*. Maun: University of Botswana.

National CBNRM Forum (2001) *Proceedings of the Second National CBNRM Conference in Botswana and CBNRM Status Report*. Gaborone Sun, 14–16 November 2001, Gaborone, Botswana. Gaborone: IUCN Botswana.

Ngami Times (2001) Bigger planes may use local airports. The Ngami Times, 9–10 March.

Oppermann, M. and Chon, K.S. (1997) *Tourism in Developing Countries*. London: International Thomson Business Press.

Pantin, D.A. (1998) *Tourism in St. Lucia*. St. Augustine: Sustainable Economic Development Unit for Small and Island Development States, University of the West Indies.

Perkins, J.S. and Ringrose, S.M. (1996) *Development Co-operation's Objectives and the Beef Protocol: The Case of Botswana, A Study of Livestock/Wildlife/Tourism/Degradation Linkages*. Gaborone: Department of Environmental Science, University of Botswana.

Plog, S.C. (1974) Why destination areas rise and fall in popularity. *Cornell Hotel and Restaurant Quarterly* 14 (4), 55–58.

Prosser, R. (1994) Societal change and the growth in alternative tourism. In E. Cater and G. Lowman (eds) *Ecotourism: A Sustainable Option* (pp. 19–37). New York: Wiley.

Ratz, T. (2002) Residents' perceptions of the socio-cultural impacts of tourism at Lake Balaton, Hungary. In G. Richards and D. Hall (eds) *Tourism and Sustainable Community Development* (pp. 36–47). London: Routledge.

Scott Wilson Consultants (2001) *Integrated Programme for the Eradication of Tsetse and Tyrpanosomiasis from Ngamiland: Environmental Impact Assessment, Draft Final Report*. Edinburgh: Scott Wilson Resource Consultants.

Seckelmann, A. (2002) Domestic tourism – a chance for regional development in Turkey? *Tourism Management* 23 (1), 85–92.

Stellenburg, P. (2003) Prostitution and tourism in the Okavango Delta: The case of Maun. Unpublished paper, Maun.

Terry, M. E. (1999) The economic and social significance of the handicraft industry in Botswana. PhD thesis, School of Oriental and African Studies, University of London, London.

Tosun, C. (2001) Challenges of sustainable tourism development in the developing world: The case of Turkey. *Tourism Management* 22, 289–303.

WCED (World Commission on Environment and Development) (1987) *Our Common Future*. London: Oxford University Press.

Wood, R.E. (1997) Tourism and underdevelopment in southeast Asia. *Journal of Contemporary Asia* 9, 274–287.

World Travel & Tourism Council (2007) *Botswana: The Impact of Travel and Tourism on Jobs and the Economy*. London: World Travel & Tourism Council.

Chapter 13

Sustainable Tourism on Commonages: An Alternative to Traditional Agricultural-based Land Reform in Namaqualand, South Africa

SHARMLA GOVENDER-VAN WYK and DEON WILSON

Introduction

Since 1994, the South African government has developed two strategic policies that embrace the principles of sustainable development: Tourism and Land Reform. Both policies seek redress and economic development for previously disadvantaged black people. In terms of the land redistribution programme (as one leg of the land reform programme), the commonage sub-programme has primarily supported an agrarian style development. A commonage is essentially land set aside for communal agricultural usage but owned by a government body and in South Africa's case owned by municipalities. By promoting one development option, other livelihood opportunities such as tourism have not been explored. Yet, tourism has been recognised by the South African government as a priority sector for national economic growth and development.

It is widely accepted that sustainability is one of the most important challenges that the tourism industry faces. Weaver and Lawton (2000) note that, in the past, the focus on sustainable development has tended to concentrate on conventional economic activities such as agriculture, mining, forestry, fisheries and manufacturing, to the exclusion of the tourism industry.

This study demonstrates that rural people, who have access to commonage land, do not have to depend on farming as the only mode of survival and that tourism ventures could also provide sustainable livelihoods. In terms of the primary data collection, the following projects in the province of the Northern Cape were selected, based on judgement sampling techniques. Judgement sampling is a form of purposive sampling that uses the judgement of an expert in selecting cases or it selects cases with a specific purpose in mind:

- six commonage projects initiated by the Department of Land Affairs (DLA) in three Namaqualand regions (Steinkopf, Springbok and Richtersveld); and
- a sustainable tourism venture (Eksteenfontein) that had been instigated by the Richtersveld community.

Once the projects were selected, key role players in the projects (municipal and DLA officials) aided in purposively selecting users for the interviews. Thirty-four face-to-face interviews (out of a possible 66 users) were conducted with commonage users from the six commonage projects over a 10-day period in November 2004. Four officials, one each from the Departments of Agriculture and Land Affairs, Nama Khoi Municipality and Richtersveld Municipality, were also interviewed.

The sustainable tourism venture is a pioneering project at Eksteenfontein in the Richtersveld in Namaqualand. Forty-two face-to-face interviews were conducted with adult (18 years and older) members (out of a possible 200) of the Eksteenfontein community over nine days in November 2004.

The results of the study are presented in two sections:

- The first section provides an overview of the South African Land Reform Programme as an engine of growth and looks at empirical evidence on the sustainability of land reform as demonstrated through the commonage sub-programme in Namaqualand.
- The second section deals with the paradigm of sustainable tourism as a livelihoods option for commonage users, through an examination of the sustainable tourism conservancy in the Richtersveld in Namaqualand.

Overview of Land Reform Policy in South Africa Since 1994

A three-pronged market-assisted land reform programme was launched in 1994, aiming at tenure reform, restitution and land redistribution (Ramutsindela, 2003 and Chapter 10 in this book).

The tenure reform programme sought to validate and harmonise forms of land ownership that evolved during colonialism and apartheid. It was an attempt to redress the dual system of land tenure where whites owned land as private property as opposed to communal land allocation among blacks (Ramutsindela, 2003). The majority of rural blacks lived on communal land that is registered as the property of the State under the erstwhile South African Development Trust, with tribal chiefs installed as custodians of communal land (Department of Land Affairs, 2003a).

Land restitution forms the second pillar of the land reform programme. It aimed to redress the imbalances in land ownership that had been by such legislation as the infamous Natives Land Act, 1913 (Act No. 39 of 1913). The type of restitution is determined by effects of land dispossession, namely, landlessness, inadequate compensation for the value of the property and hardships that cannot be measured in financial or material terms (Department of Land Affairs, 1997) (Box 13.1).

Box 13.1   Community Perceptions and Participation in the Management of the Isamangoliso Wetland Park, South Africa: Towards a Sustainable Tourism Development Initiative

LINDISIZWE M. MAGI and THANDI A. NZAMA

Currently South Africa has eight World Heritage sites, two of which are located in KwaZulu-Natal. These two are Isimangaliso Wetland Park (1999) and Ukhahlamba-Drakensberg National Park (2000). The main focus of this case study is to determine the degree to which Isimangaliso Wetland Park communities participate in activities that promote sustainable ecotourism development as well as reveal the existing ecotourism management practices in the study area Isimangaliso Wetland Park (known formerly as Greater St Lucia Wetland Park – received a new name Isimangaliso Wetland on 1 November 2007 with an intention of placing emphasis on its unique African identity).

The Park was the first in South Africa to receive the World Heritage status in July 1999 because of its unique ecological processes, superlative natural phenomenon and scenic beauty, exceptional biodiversity and a large number of threatened and endangered species. Currently, it is South Africa's third largest park, spanning 280 km of Indian Ocean coastline. The challenge is to retain its World Heritage status by protecting and maintaining its ecological integrity. To protect the Park from degradation by tourist activities and local communities, emphasis has been placed on creating a balance between utilisation and protection of the resource.

Isimangaliso as an Ecotourism Resource

Isimangaliso Wetland Park is perceived as an important ecotourism resource and this assertion is supported by the decision that was taken in 1996 when conservation and ecotourism were deemed to be the best land use option for the Park (Zaloumis, 2007). Ecotourism as a strategy for sustainable development and conservation of physical and socio-cultural environment demands that the 'local community' should be involved in planning, development and operation which should contribute to their well-being (Ceballos-Lascurain, 1996). As an ecotourism resource, Isimangaliso Wetland Park has a mandate to protect its World Heritage values while supporting economic development of local communities, restitution and social justice (Zaloumis, 2007). Unfortunately, like many parks in South Africa, Isimangaliso's biological and cultural diversity are threatened by unsustainable land-use practices. The challenge for the Park Authority is therefore to maintain the equilibrium between the activities that promote ecological processes to operate as freely as possible in order to protect the integrity of the ecosystems (biocentrism) and those that promote use and maximisation of direct human consumption (anthropocentrism) (Cooper *et al.*, 2000; see Magi, 2007).

Community Involvement in the Management

Isimangaliso Wetland Park has several communities living within its boundary. These communities were removed by the government of the day, when the land was proclaimed a game reserve in 1895. With the advent of a new democratic order, these communities applied for land restitution. In 2007, after extensive negotiations, nine local communities had 75% of their land claims settled. The title deeds were passed from the state to the new land owners. This settlement coincided with the decision of the Park Authority to give local communities first priority to ensure that they are involved in the activities taking place in the Park.

The official policy of Isimangaliso Wetland Park as a World Heritage Site is to seek activities that promote sustainable ecotourism development. It also recognises that ecotourism, as a strategy for sustainable development and conservation of outdoor recreation landscapes, demands that the local community should participate in the planning, development and management of the hydro-geopark (see WTO, 2004). Indeed, the notion of community involvement in ecotourism development and decision-making is one, which is advocated for by the Tourism White Paper (Department of Environmental Affairs and Tourism, 1996) in South Africa. To evaluate the involvement of the local community in the management of the Park, a case study with a total of 303 responses was conducted. The majority of the Park officials (65%) indicated that there was high involvement of the local community. On the other hand, the majority of tourists (58%), the tour operators (65%) and the local community (68%) indicated that community involvement in the management of the Park was significantly low. Surprisingly, only a small number of all these groups suggested that there was high level of community involvement in the management of the Park. Among local people especially those who support their livelihood by selling artefacts to the tourists have expressed that, even though they are at times consulted, their welfare is not the top priority of the custodians of the Park. These concerns were expressed after the implementation of the ban of 4 × 4 vehicles at the Isimangaliso beach area. The concerns were also expressed notwithstanding the attainment of the co-management agreements with some local tribal authorities, such as for example the Mbuyazi Local Tribal Council.

A commonly held view is that participation benefits in ecotourism management practices enhance local economic development (Aronsson, 2000). In the case study undertaken in the Wetland Park, the majority of tour operators and community members (all below 50%) perceived most of the practices to be less beneficial for ecotourism development in the area. Similarly, the practices of tourism safety and security as well as the Park and estuary preservation were seen by the community and tour operators to be relatively beneficial for local development initiatives. The practices of land and resource acquisition were perceived by a minority of the community and tour operators to be less beneficial for local development systems in the study area.

Notwithstanding the negative view of the local community regarding participation in ecotourism development, there are several joint activities where stakeholders are making a contribution to sustainable economic development of the area. Some of the projects which have achieved success in the area include

- Establishment of co-management practices, where there has been settlement of nine major land claims.
- Wetland rehabilitation, dune rehabilitation and alien plant rehabilitation.
- The extension of the Park through the incorporation of private and communal land, which would improve the viability of ecotourism activities.
- Preparing an Integrated Management Plan which includes the re-introduction of endemic game, such as elephants, zoning, setting carrying capacity systems and so on.

- Forming community, private and public partnerships towards growing the tourism development initiatives, such as community accommodation, and so on.
- Job creation and employment opportunities for the benefit of the local community.

The above-mentioned factors are an endeavour by the Wetland Authority to ensure that local communities benefit from the hydro-geopark and that the World Heritage status of the Park is maintained.

Conclusions

Notwithstanding the fact that the declaration of the Isimangaliso Wetland Park as a World Heritage Site has created more popularity and economic viability of the area, it has also come to be recognised for its cultural and heritage attributes. The relationship between the local communities and Park Authorities is getting stronger as the two main stakeholders get into co-management agreements, which are affording the communities greater participation and ownership of the hydro-geopark.

According to Zaloumis (2007), since the establishment of the Park Authority in 2000, local people have been given priority in business opportunities, jobs, training and development projects. The involvement of local people is in line with the vision of both the government and the local communities that seek to promote ecotourism and nature conservation. The main challenge of Isimangaliso and its conservation partner Ezemvelo KZN (KwaZulu-Natal) Wildlife is to explore how social justice and restitution of past wrongs can be corrected through management of protected area while still meeting the nature conservation objectives. One of the major objectives of the Wetland Authority is to ensure that the World Heritage Site is developed in a way that ensures that local residents benefit from the Park.

The full article is available on request from the case box authors.

Land redistribution was conceived as a means of opening up the productive land for residential and agricultural development. The government set itself a target of redistributing 30% of the country's commercial agricultural land (about 24 million hectares) over a five-year period from 1994 to 1999 (Department of Land Affairs, 1997). This target has been extended since the review of the programme in 2000 to redistribution of 30% of agricultural land by the year 2014 and encompasses all agricultural land redistributed through all three programmes (Department of Land Affairs, 2003a).

This study primarily focuses on the redistribution programme, in particular the commonage sub-programme, given that the programme has resulted in substantial land transfers in the Northern Cape, primarily in the Namaqualand region.

The Commonage Sub-programme

A commonage can be defined as follows: 'commonage or common pasture lands are lands adjoining a town or village over which the inhabitants of such town or village either have a servitude of grazing for their stock and more rarely, the right to cultivate a certain portion of such lands, or in respect of which the inhabitants have conferred upon them by regulation certain grazing rights' (Dönges & Van Winsen, 1953: 303).

Approximately 150 commonage projects have been implemented by the DLA since 1997, and all of them are grazing projects or small-scale crop projects (Department of Land Affairs, 2004). The policy, however, allows the DLA 'to ensure that commonage land needed by previously disadvantaged communities for agricultural *and other entrepreneurial business purposes* is made available for such purposes' (Department of Land Affairs, 2000: 8). The commonage sub-programme proclaims to act as a nursery to small and emergent farmers so that entry to another land reform sub-programme that primarily caters to the development of black commercial farmers is possible. This sub-programme, called the Land Redistribution for Agricultural Development Sub-programme, or LRAD, also falls under the Redistribution Programme. LRAD offers subsidies on an individual basis to qualifying applicants to purchase agricultural land from white farmers. The subsidies range from R20,000 to R100,000 and are based on own contribution in kind, labour and/or cash. However, this linkage is yet to be established and only one known case of 'graduation' from subsistence commonage user to black commercial farmer exists.

Although the government is chasing its target of redistributing 30% of commercial agricultural land by 2014, the question of the sustainability of land reform projects and its contribution to the socio-economic growth of rural people remains to be answered.

Sustainable Livelihoods and Land Redistribution

One of the most damning criticisms of the land redistribution efforts in South Africa is that it is executed in isolation of other livelihood strategies and economic sectors. Among rural farming communities, many derive a part-livelihood from farming, a part from migrant labour/mining and a part from other activities such as arts and crafts. It has been observed that there is a close correlation between the diverse modes of livelihood and the idea of diversification and sustainability of livelihoods over time among farming communities. Bryceson (1999) contends that 60–80% of rural household income in sub-Saharan Africa in the late 1990s was derived from non-farming sources. Such trends have led to the coining of the term 'sustainable livelihood'. A livelihood is considered to be sustainable when it can survive stresses and shocks and maintain or enhance its capabilities and assets both now and in the future while not undermining the natural resource base. Pasteur (2001) contends that an analysis of livelihoods involves identifying and understanding the assets and options available to poor people and the vulnerable context within which they operate.

Since the emergence of land reform in southern Africa during the 1980s (South Africa and Namibia in the 1990s), the question of sustainable land reform has plagued development planners. The South African government has made several attempts to develop a feasible development strategy. In 2003, the DLA published a framework for accelerating land reform for 'sustainable development'. This framework recognised the importance of the implementation of a sustainable land reform programme that is dependent on an integrated approach to land reform in close collaboration with key government and non-governmental stakeholders (Department of Land Affairs, 2003b). A think tank on land reform in southern Africa held in 2003 revealed that there was a general incongruity between land policy and rural development, with government pursuing a compensatory (30% target) rights-based approach to land reform rather than a sustainable development approach (Human Sciences Research Council, 2003).

Sustainable development clearly embraces the environment, people and economic systems (Hunter, 1997; Murphy, 1995; Swarbrooke, 1999). Table 13.1 presents an adaptation of some of Murphy's (1995) components for sustainable development based on the Bruntland Report in 1987 for the World Commission on Environment and Development entitled *Our Common Future*. These components are examined against the current land reform policy in South Africa to obtain an overview of its sustainability factors.

The DLA Commonage Sub-programme in Namaqualand

The DLA has adopted a more developmental approach especially in the Northern Cape through the commonage sub-programme. About 307,000 ha of agricultural land in Namaqualand were purchased through this sub-programme to add to the existing municipal commonage for use by poor residents, mainly for grazing and small-scale agricultural production (Department of Land Affairs, 2004). In the study area a total of 26 farms, in extent of 111,000 ha, were purchased to make up six commonages for subsistence and emergent livestock farmers in the three towns (Figure 13.1). The farms were purchased after the communities in the three towns have

Table 13.1 Comparing the main components of sustainable development with current land redistribution policy and implementation. Table shows only some factors crucial to land reform

Sustainable development component	*South Africa*
Setting ecological limits and equitable standards	Environmental guidelines exist but not integrated into the planning processes
Redistribution of economic activity and reallocation of resources	Three million hectares of land redistributed instead of 12 million hectares as at 2005
Conservation of basic resources	Environmental guidelines ignored
Community control	Limited
Broad national/international policy framework	Lack of integration of planning for land redistribution with other sustainable development initiatives
Economic viability	The government provides grants to kick-start farming operations but because of the limited grant size, many farming operations have not been economically viable

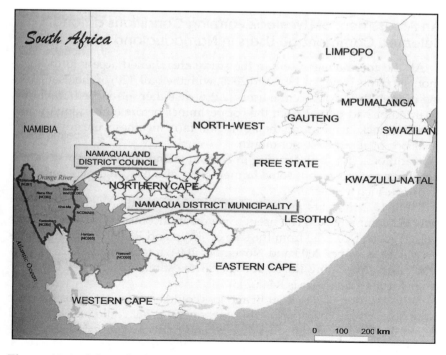

Figure 13.1 Map of selected commonage projects

expressed the need for land for livestock farming following mass retrenchments in the copper and diamond mining industries in Namaqualand.

Hoffman and Rohde (2000) claim that national land redistribution policies are not effective in Namaqualand because land prices are high and private land ownership is almost impossible, therefore commonage has been the principal mode of land reform implementation in this part of the province. In addition, grazing and agricultural lands can be considered marginal in Namaqualand where vast tracts are showing signs of overgrazing and land degradation.

The Centre for Arid Zones Study in the United Kingdom also noted that livestock farming in Namaqualand does not have a sustainable future (Young, 2002: 1) and one of the crucial reasons behind this theory is the lack of farm diversification strategies and the reluctance of people to let go of traditional livestock farming practices in the area. Young (2002) advocates conservancy development coupled with tourism as a possible livelihood strategy for some of Namaqualand's communities.

Other livelihoods in Namaqualand have also not fared so well. The region has relied heavily on the mining sector but first the copper reserves and now the land-based diamond deposits are exhausted. Large-scale decommissioning of mine workers left many more families without incomes.

An Assessment of Livestock Farming Conditions among Selected Commonage Users in Namaqualand

All the commonage users in the study area gained access to the commonage farms between 1998 and 2004, with the bulk (20) of the users coming in from 2001. Most of the users (22) were either retrenched, medically boarded or had retired from the copper mines before embarking on full-time livestock farming and their only non-farm incomes were government pensions of about R740 per month.

Traditional livestock farming in Namaqualand involves primarily cattle, sheep and goats, but some farmers have also acquired pigs and small livestock such as chickens.

Table 13.2 presents an outline of livestock farming activities on the selected commonages. The users received an income of R13,475 from the sale of goats, R207,750 from the sale of sheep and R8400 from cattle sales between December 2003 and November 2004, a total of R229,625 in livestock sales, averaging R6753 per user. The estimated turnover for all 66 users for the given year is R445,743.

No estimation of actual profit and loss could be determined, as the costs were not factored in as part of the assessment. What could be ascertained is that the users pay transport costs to and from the commonages; medicine for the stock and food and water for the stock. Stock numbers vary and the sales averages will differ for each of the farmers, with only few of

Table 13.2 Livestock farming on commonages

Type of animal owned	Total number	Number sold in last 12 months	Average[a] selling price per unit	Total Number sold	Number slaughtered for automatic consumption
Goats	455	49	R275	R13,475	100
Sheep	2825	554	R375	R207,750	123
Cattle	670	6	R1400	R8400	
Pigs	11	0	0		
Chicken	72	0	0		36
Total	4033	609	–	R229,625	259

[a]The Department of Agriculture was asked to verify the average prices.

the farmers actually showing profits. Only four of the users sold the skins and milk to earn extra income. The market for goat meat is not profitable, and is for own consumption rather than sales.

The users were then questioned on the advantages and disadvantages of livestock farming on the commonages (Table 13.3). The numbers in brackets next to each issue indicate the number of responses received. While the disadvantages outweigh the advantages by far, the majority of the users (90%) did indicate that the drought, that has plagued the area from 2002, played a major role in their negativity towards livestock farming and that a rainy season would result in more profits. A majority (30) of the users agreed that farming conditions were conducive for livestock farming with proper management and good rains but the current drought conditions were hampering livestock production. Plate 13.1 illustrates some of these conditions on two of the sampled commonage farms.

Quality of Life since Participating in the Commonage Programme

Table 13.4 provides an overview on whether access to commonages has resulted in improving the users' lives in relation to some identified factors. The opinions of the officials from municipalities, Departments of Land Affairs and Agriculture were also factored in. Most of the respondents indicated that there were no improvements in relation to housing and moveable assets and that they could not provide education for their children from the minimal earnings obtained from livestock farming. While most of the respondents reported some improvement in terms of income, they asserted that this was only a slight difference. Respondents added

Table 13.3   Advantages and disadvantages of livestock farming on commonages (*N* = 34)

Advantages	Disadvantages
Improves household income (10)	Commonage land is far from town and home (35–60 km) (30)
Free advice from surrounding white commercial farmers (14)	Drought and no drought relief
Grazing for animals (34)	Wild animals/predators such as jackal and 'witkruis' eagle that are protected (25)
No expansion of livestock after accessing the commonages (10)	Brackish water and limited grazing fields (34)
Some farmers have sole use of some of the 26 farms that were purchased for these six commonages (10)	Few boreholes on commonages (25)
Improves household food consumption (34)	Infrastructure on some of the farms is in poor condition such as the fencing and wind pumps (25)
	Soil erosion (30)
	Farms are divided into summer and winter farms and farmers felt that this disadvantaged them for six months of a year (30)
	Poor rotational grazing practices (34)
	No training or additional subsidies provided to farmers and therefore cannot expand livestock farming enterprise as they would like to (30)
	Livestock restrictions (34)

Table 13.4   Improvement/non-improvement of quality of life (*N* = 38)

Factors	Improvement	No improvement
Land access	38	
Housing	15	23
Farming, for example, increase in livestock	30	8
Education of children	14	24
Food	34	4
Income	24	14
Other moveable assets	10	28

Nanasan (part of Porth Nolloth commonage)

Spiönkop (part of Springbok commonage)

Taaibosmond (Steinkopf)

Plate 13.1   Nanasan, Spiönkop and Taaibosmond give an indication of arid conditions on the commonage farms

that the money gained from livestock farming was often reinvested in the business, either to buy food or medicines for the livestock. While all the respondents experienced improvements in terms of access to land, they believed that land ownership rather than land access would have provided a tangible benefit to them. However, this would go against the policy commonage principle that emphasises access as opposed to ownership.

Expression of Interest for Tourism Development on Commonages

There are currently no sustainable tourism activities on the selected commonages. The farms were initially purchased from white livestock farmers and this practice has been pursued by the new users. Ten of the users stated that they had expressed an interest in tourism activities to the municipalities. They had wanted to establish guesthouses on two of the commonage farms, 4 × 4 routes, bird watching, conservation tours and wildlife and floral viewing, but the ideas were never fully accepted. The municipalities also discussed these opportunities with the users but half of the livestock farmers were afraid to venture out of their traditional livelihoods mode. The others who replied negatively asserted that the main reason for the lack of interest in establishing tourism ventures on the commonages was because there was no subsidisation of these activities, and also that they did not have the skills to start and/or sustain such activities.

The commonage case studies were evaluated on the basis of whether livestock farming has been a successful form of livelihood on the commonages. Respondents indicated that their livestock farming enterprises were barely successful. They pointed out that farming conditions were far from ideal and that lack of access to water and poor infrastructure were the key reasons for this. The study further revealed that, in relation to tourism, the commonage users overwhelmingly support ecotourism and/ or nature-based tourism ventures on the commonages. The next section provides an overview of the concept of sustainable tourism and investigates a sustainable tourism venture at Eksteenfontein in the Richtersveld in Namaqualand.

Sustainable Tourism

The concept of sustainable tourism is a very broad, imprecise developmental concept and it is not the intention of this chapter to posit a definition but to harness the broad principles and relate this to land reform on commonages. The study supports the definition contained in the White Paper on Tourism (Department of Environmental Affairs and Tourism, 1996: 3) that defines sustainable tourism as 'tourism development, management and any other tourism activity which optimise the economic and other societal benefits available in the present without jeopardising the potential for similar benefits in the future' (see also Saarinen, Chapter 5 in this book). This definition can also be used to describe the broad policy intent of the land reform agenda, which is to ensure that the targeted people use the natural resource (land) efficiently and for social and economic development.

Despite the sustainable tourism vision promoted in the White Paper on Tourism, disadvantaged communities and population groups remain highly marginalised from the 'mainstream' tourism industry and the national, high-profile initiatives that underpin its notable growth.

Both the national priorities for reform and tourism are seeking redress and economic development for the previously disadvantaged communities of South Africa and both accentuate the sustainability issues. These are laudable but not easy targets that become even more difficult to attain when policies are constructed and implemented in silos. The following section provides an assessment of a sustainable tourism venture in the Richtersveld area of Namaqualand.

Rooiberg Tourism Conservancy Project at Eksteenfontein in the Richtersveld

The Richtersveld in Namaqualand consists of four villages, Kuboes and Sanddrif in the North and Eksteenfontein and Lekkersing in the South. The Richtersveld forms part of Little Namaqualand. The original inhabitants of the Namaqualand were Khoi Khoi, but also included some San people. They were present in the area long before Dutch colonisation of the Cape.

After unification and during 1925, the South African government appointed a commission to investigate the legal position of the Richtersveld. In 1925, diamonds were discovered near Port Nolloth and many people moved into the area (Boonzaaier *et al.*, 1996). In 1930, the Minister of Lands issued a certificate of reservation in respect of the Richtersveld Reserve land under the disposal of the Crown Lands Act in favour of the Minister of Native Affairs for the use of the persons residing therein (Land Claims Court, 2001). However, certain pieces of land such as diamond-rich areas were excluded from the reservation and this exclusion became the subject of the long-running land restitution court case between the Richtersveld communities and Alexkor Limited (Boonzaaier *et al.*, 1996).

In 1998, a land claim for 85,000 ha of land in the Richtersveld (including the diamond-rich land that belongs to Alexkor) was filed with the Land Claims Court by the four communities that comprise the Richtersveld. The communities lost the case but in 2001 they appealed to the Constitutional Court, which decided that those communities were the legal owners of the land and considered the appeal in terms of the indigenous rights of the communities. Eventually, the Constitutional Court decided that the erstwhile apartheid government and Alexkor had unfairly dispossessed the communities of their land rights because of the mineral wealth (Strauss, 2004).

The communities registered the Richtersveld Communal Property Association (CPA) that will take possession of the land once the Minister of Land Affairs finalises the transfer of the property. In the interim, the Richtersveld Municipality is the appointed manager. The communities are also still awaiting a response in terms of the settlement/compensation package from the government and Alexkor.

Eksteenfontein is one of the 13 towns along the South–North Tourism Route (South Africa). There are approximately 800 people in Eksteen-fontein, 400 of them adults. Some of the adult members are employed on the mines and some have left the area to pursue tertiary studies or seek employment in other provinces. Eksteenfontein has two community-run guesthouses. In the centre of town is the *Kom Rus 'n Bietjie* (come rest awhile) guesthouse. After acquiring funding, the local women's association renovated an old mining shack into this guesthouse. The guesthouse is fully electrified and has the simple comforts of home such as beds, shower, bath and fully fitted kitchen. There is no television, air-conditioning or even a fan and tourists would have to deal with mosquitoes in summer. The area is, however, malaria-free and safe. While the local tourism officer contends that the bare minimum was necessary for tourists who wanted to be close to nature as possible, there is a definite need for an upgrade of the guesthouse in terms of tiling, painting and bedding. Management has mentioned that there are plans to upgrade but sourcing funding was problematic. The guestbook comments also revealed that most of the tourists found their stay quite pleasant. The village women, who were also the guesthouse managers, prepared the traditional food served, which is a unique touch.

The Eksteenfontein community has also initiated a conservancy project in 2002, the Rooiberg Conservancy project that is about 50 km from the town. The vision of the Rooiberg Conservancy Project is 'to protect and manage the unique biodiversity and natural landscape to the advantage of the local people and all of humankind' (Richtersveld Community Conservancy, 2004). The conservancy has a guesthouse and traditional Nama campsites with *matjieshuts* or mat huts. These facilities do not have electricity. The Rooiberg Conservancy Management manages these facilities but has not employed people on a full-time basis. The Eksteenfontein Tourism Office maintains the finances of both guesthouses.

The 42 people interviewed are beneficiaries of the conservancy development and were either directly or indirectly involved with the conservancy. Only 13 of the respondents were actually participating in the conservancy project and the levels of participation include management, cartography (mapping of the area), tour guides and cultural guides.

Table 13.5   Training received

Conservancy management	2
Nature conservation	2
Tourist management/tour guide	2
Cartography	2

Participants in the conservancy project had been selected on the basis of their residency in the Richtersveld and their age at least 18 years.

Only 8 of the 13 members that are involved in the conservancy project have been trained (Table 13.5). One member of the community mentioned that while some people were trained to be guides, they did not have a passion for the work and the wrong people were targeted for the job. Another person pointed out that some of the training did not coincide with implementation and people had been trained but not employed. The researcher observed that the proposed museum for the area has not opened due to a lack of funding and one person was trained but is currently unemployed.

People were also asked to comment on plans for the conservancy in relation to tourism (Table 13.6). The perceptions of the majority of the interviewees tie in with the tourism management plans for the conservancy, namely the development of 4 × 4 trails, eco-sensitive hiking trails and conservation of the flora and fauna in the area.

Table 13.6   Opinions on the plans for the conservancy

Plan	*Opinion*
To expand the guest house business	28
To develop nature conservation programmes for tourists	20
To develop a four by four (4 × 4) route for tourists	26
To protect the natural environment and animals for tourists	32
To develop campsites for tourists	24
To develop nature tours	26
To develop bird watching for tourists	15
To develop game viewing for tourists	17
To develop game hunting facilities for tourists	11
To developing eco-sensitive hiking trails for tourists	32

Table 13.7 Economic and social spin-offs from conservancy project

Economic spin-offs of conservancy development	*Social spin-offs*
Creation of the following job opportunities Signage (sign writers) Caterers Guesthouse managers Tour and cultural guides Cartographers Rangers	Reduced unemployment
Upgrading of infrastructure and use of local skills	Reduced alcoholism
Increased spending by tourists provided opportunities for the development of arts and craft and retail businesses	Increased capacity building
	More youth involvement
	More community involvement
	Educational improvements

It is estimated that the conservancy receives 80% of its tourists from South Africa and 20% from outside the country. There are generally four tourists per day, off-season between October to March, and in peak season, between April and September, there are on average 13 tourists per day. It is likely that tourists spend an average of R750 per day per tourist in Eksteenfontein, supporting the two local shops, guesthouses and going into the conservancy. Each tourist stays on average 3–5 days. The estimated annual income from the conservancy development therefore amounts to R549,000 (off-season) and R1774,500 (peak season). This excludes the rental of equipment or vehicles. Table 13.7 highlights the community's views on the spin-offs received from this venture.

Tourism Development (Present and Future)

The majority (23) of the interviewees rate tourism as very important as compared to livestock farming and/or mining while the others (19) viewed tourism as equally important to mining and livestock farming. There is a perception among some of the community members interviewed that tourism can do more harm than good, but this is a minority view.

Tourism is currently seen as the economic 'saviour' in response to the decommissioning of the mines and the difficulties of livestock farming in

Namaqualand. However, it may be idealistic to rely on tourism alone and there is a need to look at other economic activities that can be offered to community members that might not be interested in the tourism developments in the area. In relation to infrastructure development, there are plans to improve the roads, electrify areas where there is no electricity (except within the conservancy), improve the signage on the roads to the conservancy and Eksteenfontein and upgrade the guesthouses.

Interviewees stated that the conservancy tourism project could generate sustainable livelihoods for the Eksteenfontein community. There are community spin-offs and in 10 years tourism will offer full-time livelihood opportunities.

The Richtersveld CPA, as owners of the Rooiberg Conservancy, plans to outsource all the tourism businesses to the community and this will include the guesthouses, campsites, tourism office and museum. They will be asked to tender for the businesses. Community members will be encouraged to form joint ventures with non-Richtersvelders to promote investment in the area. Community members who are currently operating some of the businesses are concerned that they would not stand a chance of winning any of the tenders, but improved communication channels between the management and the community could allay these fears.

The management committee noted that not all members of the community could benefit from tourism opportunities but these realities should also be clarified to the community. Interviewees stated that the youth are growing up within a culture of tourism and they have the potential to develop tourism and sustain it. Most of the older residents do not understand tourism and how tourism could provide benefits to them because they were either miners or livestock farmers. However, these livelihoods should continue to be the options for the community and become integrated with tourism activities in the area.

Generally, there appears to be some economic and social spin-offs in relation to the conservancy. The analyses illustrate that not all members of the community are informed on the conservancy development and feel that there is nepotism with regard to training and/or job opportunities. The majority of the interviewees agree that the conservancy tourism project, if linked with other initiatives in the area such as the establishment of the transfrontier conservation area with Namibia and mining, can create sustainable livelihoods for the Eksteenfontein community.

Conclusion

The study has ascertained that livestock farming cannot provide sustainable livelihoods to people accessing commonages. While livestock farming has been the traditional livelihood generator in Namaqualand, commonage users did express interest in tourism but this was never developed into a comprehensive strategy.

In general, there appears to be positive economic and social spin-offs from the sustainable tourism venture. The study established that over a 12-month period, one sustainable tourism venture benefiting 300 adult members was more successful in generating profits than 34 micro-live-stock farming enterprises on six commonages, benefiting 34 commonage users. Hoffman and Rohde (2000) assert that livestock farming on com-monages in Namaqualand should ideally yield a net annual income of R10 per hectare but states that this is not achievable because of the poor farming conditions on the commonages.

Comparatively, it would appear that the tourism venture has a financial edge over the livestock farming enterprises (R445,743 income generated from livestock sales on the six commonages versus R549,000 (off-season) and R1 774,500 (peak season)). In terms of the tourism venture, most of the spin-offs have been positive but the benefits, that is, social, ecological and financial, must be candidly explained to the communities who are involved in such initiatives.

The Eksteenfontein case study has demonstrated that communities can create sustainable livelihoods through tourism on commonages but only if the following issues are prioritised:

- Open communication channels between the community and the management of such ventures to minimise misunderstandings and involve the community in all levels of decision-making.
- A comprehensive skills development plan that would not only emphasise the type of skills needed for the venture, but also who should be targeted for training and when skills training should com-mence. Skills development should ideally coincide with the period of employment.
- The development of a sustainable tourism strategy that fits in with the development vision of the area.
- The development of a monitoring and control system to identify weaknesses in the venture and capitalise on the strengths.

Acknowledgement

The primary data collection was made possible through funding from the Southern African Labour Development Research Unit based at the Centre for Social Science Research, University of Cape Town.

References

Aronsson, L. (2000) *The Development of Sustainable Tourism.* London: British Library Cataloguing-in-Publishing Data.
Boonzaaier, E., Berens, P., Malherbe, C. and Smith, A. (1996) *The Cape Herders.* Cape Town: David Philip.

Bruntland, G. (ed.) (1987) *Our Common Future: The World Commission on Environment and Development.* Oxford: Oxford University Press.

Bryceson, D.F. (1999) *Sub-Saharan Africa Betwixt and Between: Rural Livelihoods Practice and Policies* (Working Paper No. 43). Afrika-Studiecentrum.

Ceballos-Lascurain, H. (1996) *Tourism, Ecotourism and Protected Areas.* Gland: International Union for Conservation of Nature and Natural Resources.

Cooper, C., Fletcher, J., Gilbert, D. and Shepherd, R. (2000) *Tourism: Principles and Practice.* New York: Longman Publishing.

Department of Environmental Affairs and Tourism (1996) *White Paper on Tourism.* Pretoria: Department of Environmental Affairs and Tourism.

Department of Land Affairs (1997) *White Paper on South African Land Policy.* Pretoria: Department of Land Affairs.

Department of Land Affairs (2000) *Commonage Manual.* Pretoria: Department of Land Affairs.

Department of Land Affairs (2003a) *Review of the Land Redistribution for Agricultural Development Sub-programme.* Pretoria: Department of Land Affairs.

Department of Land Affairs (2003b) *Accelerated Land Delivery for Sustainable Development: Strategic Management Framework for the Branch: Land and Tenure Reform.* Pretoria: Department of Land Affairs.

Department of Land Affairs (2004) *Programme Performance Draft Report: Redistribution and Tenure Programmes of the Department of Land Affairs.* Pretoria: Department of Land Affairs.

Dönges, T. and Van Winsen, L. de V. (1953) *Municipal Law.* Cape Town: Juta.

Hoffman, T. and Rohde, R. (2000) *Policy and its Influence on the Sustainable Development of Paulshoek.* On WWW at http://www.maposda.net/Global/reports/lesotho/Session4a_South%20Africa.pdf. Accessed 15.9.2004.

Human Sciences Research Council (2003) *Seeking Ways Out of the Impasse on Land Reform in Southern Africa: Notes from an Informal 'Think Tank' Meeting.* Unpublished paper, Manhattan Hotel, Pretoria, South Africa.

Hunter, C. (1997) Sustainable Tourism as an Adaptive Paradigm. *Annals of Tourism Research* 24 (4), 850–867.

Land Claims Court of South Africa (2001) In the case between the Richtersveld Community, the Kuboes Community, the Sanddrift Community, the Lekkersing Community, the Eksteenfontein Community, the Adult Members of the Richtersveld Community and Alexkor Limited (First Defendant) and the Government of South Africa (Second Defendant). Case Number: LCC 152/98.

Magi, L.M. (2007) To recreate or conserve, that is the question: The case of lake St Lucia. *Journal of Tourism and Hospitality* 4 (2), 1–12.

Murphy, P.E. (1995) Tourism and sustainable development. In W.F. Theobald (ed) *Global Tourism: The Next Decade.* Oxford: Butterworth-Heinemann.

Pasteur, K. (2001) *Tools for Sustainable Development.* On WWW at http://www.livelihoods.org. Accessed 4.4.2003.

Richtersveld Community Conservancy (2004) Financial Management Plan. Unpublished manuscript, Richtersveld.

Ramutsindela, M.F. (2003) The perfect way to ending a painful past? Makuleke land deal in South Africa. *Geoforum* 33, 15–24.

Strauss, F. (2004) Die Grondvraag is die grondvraag: Grond en demokrasie in Suid-Afrika. On WWW at http://www.litnet.co.za/seminaar/floorss. Accessed 11.12.2004.

Swarbrooke, J. (1999) *Sustainable Tourism Management.* Wallingford: CABI Publishing.

Weaver, D.B. and Lawton, L. (2000) Sustainable tourism: A critical analysis in cooperative research. Research Centre for sustainable Tourism Research Report Series, Research Report 1, Australia.

WTO (World Tourism Organisation) (2004) On WWW at http:/www.world-tourism.org/frameset/frame sustainable.htlm. Accessed 14.10.2006.

Young, E.M. (2002) A sustainable future for southern Africa's rangelands? *New Agriculturalist Development* 2 (2), 1–5.

Zaloumis, A. (2007) A name for a miracle. *Isimangaliso News* 1 (1), November 2007 to January 2008, 1–10.

Chapter 14

Local Food as a Key Element of Sustainable Tourism Competitiveness

GERRIE E. DU RAND and ERNIE HEATH

Introduction

According to the World Tourism Organisation (WTO, 1999) sustainable tourism development, including food tourism should meet the needs of present tourists and host regions while protecting and enhancing opportunities for the future. The development of sustainable tourism stipulates the following requirements:

- *Tourist resources* should be preserved in such a way that allows them to be used in the future, while benefiting today's society.
- The planning and management of *tourist development* should be conducted in a way that avoids triggering serious ecological or socio-cultural problems in the region concerned.
- The overall quality of the environment in the *tourist region* should be preserved and, if necessary, improved.
- The level of *tourist satisfaction* should be maintained to ensure that destinations continue to be attractive and retain their commercial potential.
- *Tourism* should largely benefit all members of the society (WTO, 1999).

Based on the above, all tourism activities should strive to comply with this definition irrespective of which market segments they serve. This is important for food tourism, which relies heavily on the tourism infrastructure, cultural heritage and agriculture of a destination (Canadian Tourism Commission, 2003). It is therefore imperative that food tourism, which can be regarded as a new product, meets these requirements so as to contribute to sustainable tourism development.

Local food holds great potential to contribute to sustainability in tourism by: enhancing the local and regional tourism resource base; adding value to the authenticity of the destination; strengthening the local economy (both from a tourism and an agricultural perspective) and providing for environmentally friendly infrastructure. Food tourism can be regarded as a consumptive activity (berry picking, fishing, hunting, etc.) when properly managed and marketed ought to achieve the goal of sustainable tourism development. Food tourism should be environmentally friendly as it will pertain to the sustainable use of food and beverage products that are both renewable resources (Boyne *et al.*, 2002).

An Overview of Food Tourism

A review of relevant literature, current trends and best practices was conducted to determine and compare the knowledge and perspectives of experts in the areas of food tourism, destination marketing and destination competitiveness. The position of food in the tourism field and how to market it as a form of niche tourism was also established.

Food tourism has ceased to be only concerned with the provision of food for tourists in restaurants, hotels and resorts. It is the tourist that now travels in order to search for, and enjoy prepared food and drink (Hall, 2003). Food is considered as 'an expression of a society and its way of life' (Kaspar, 1986: 14), which is verified by Boniface (2003), who regards culture, both past and present, as an inevitable part of food tourism. Long (1998) accentuates the fact that food tourism is a sensory experience utilising all the senses making it central to the tourism experience. Hall (2003: xxiii) summarises it and contends that food tourism is increasingly

- recognised as part of the local culture, consumed by tourists;
- an element of regional tourism promotion;
- a component of local agricultural and economic development;
- a key element to competitive destination marketing;
- an indicator of globalisation and localisation; and
- a product and service consumed by tourists with definite preferences and consumption patterns.

A focus on how food can contribute to tourism marketing strategies is becoming more urgent and apparent in the present-day research. The roots of food tourism lie in agriculture, culture and tourism (Boniface, 2003; Cusack, 2000; Hjalager & Corigliano, 2000; Selwood, 2003; Wagner, 2001; Wolf, 2002). All three components offer opportunities and activities to market and position food tourism as an attraction and experience in a destination. Agriculture provides the product, namely food; culture provides the history and authenticity and tourism provides the infrastructure and services and combines the three components into the food tourism experience. These three components form the basis for the positioning of food tourism as one of the components in the tourism paradigm.

Food Tourism and Sustainable Destination Competitiveness

Food is seldom the key reason for visiting a destination and is often considered as part of the overall destination experience (Hjalager & Richards, 2002b; Long, 2003; Selwood, 2003). Food is becoming one of the most important attractions as tourists seek new and authentic experiences and alternative forms of tourism (Boyne *et al.*, 2003; Crouch & Ritchie, 1999b; Hjalager, 2002; Hjalager & Richards, 2002a; Selwood, 2003). It is this very paradox that is creating the opportunity for food tourism to become an important and appealing attraction in a destination. The destination can enhance the appeal of its resources and attractions by marketing them correctly (Crouch & Ritchie, 1999b). This would include product development, packaging, positioning and the promotion of the attraction.

From the preceding perspectives it is apparent that destination marketing and food tourism are interlinked. No destination can afford to ignore the importance of food as either a key or a supportive attraction.

The contribution of food tourism to the sustainable competitiveness of a destination entails the identification, development and implementation of food tourism enhancers to achieve destination competitiveness. The concept of sustainable competitiveness adopted in this study is that of Ritchie and Crouch (2003a) which entails the ability to increase tourism expenditure by attracting a larger number of tourists, providing them with satisfying, memorable experiences, profitably, while enhancing the well-being of destination residents and preserving the natural capital of the destination for future generations. Sustainable competitiveness of the destination is therefore of prime concern.

Food tourism is an amalgam of natural features, culture, services, infrastructure, access, attitudes toward tourists and uniqueness. According to Quan and Wang (2003), food tourism holds several implications such as adding value to agricultural products, providing a theme to build up attractions, utilising culture of foods as a food-related event, incorporating food into mega events and enhancing the local identity for destination marketing and development; local foods should not be regarded as trivial and ignored in tourism marketing.

The destination competitiveness framework of Ritchie and Crouch (2003a), Figure 14.1, provides a useful point of departure and frame of reference to contextualise the position, role, importance and contribution of food tourism regarding the enhancement of the competitiveness and sustainability in a destination. According to Ritchie and Crouch (2000), the most important product in tourism is the destination experience. Destination attractions and experiences they offer are increasingly regarded as key

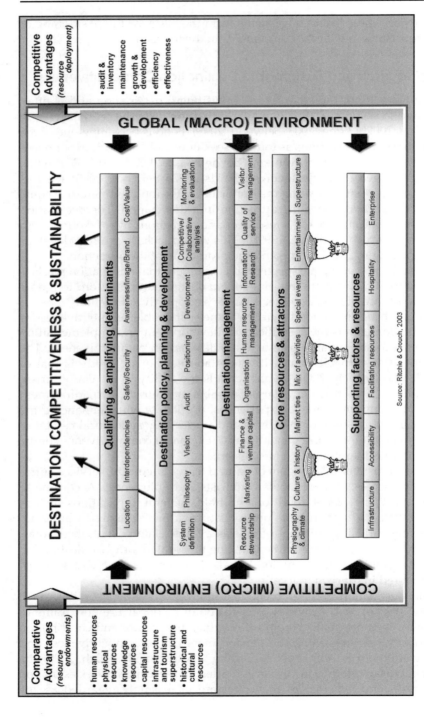

Figure 14.1 Destination competitiveness framework (Ritchie & Crouch, 2003)

elements of competitiveness and are receiving more attention and promotional funding (Ritchie & Crouch, 2000).

Destination marketing, as one of the components of destination competitiveness and sustainability, provides opportunities to achieve and ensure destination competitiveness (Crouch & Ritchie, 1999a). Changing consumer preferences, and the fact that the WTO regards Africa as an undeveloped market, provide sufficient reason to include food tourism as an attraction in destination marketing strategies. The development and enhancement of food tourism as a destination attraction by the various tourism stakeholders in South Africa could help to improve sustainable destination competitiveness. This could result in tourism destination communities receiving social and economic benefits, and tourists and visitors receiving more benefits from appropriately developed and marketed attractions (Yoon, 2002).

The Benefits and Impacts of Local Food Tourism

Local food as a component of food tourism holds great potential to contribute to sustainable competitiveness in a destination, both from a tourism development and from a destination marketing perspective. The promotion of local food is an effective way of supporting and strengthening the tourism and agricultural sectors of local economies by preserving culinary heritage and adding value to the authenticity of the destination, by broadening and enhancing the local and regional tourism resource base and by stimulating agricultural production. The development of a framework and guidelines for developing and implementing food tourism can enable destination marketers and entrepreneurs to optimise the tourism potential of local food. Boyne *et al.* (2003) reinforce the need for a framework to enable the stakeholders to cooperate and achieve the effective implementation of marketing strategies regarding food tourism. Figure 14.2 provides an outline of the process and interrelatedness between a sustainable and competitive destination, food tourism and destination marketing management.

An analysis of the relevant tourism literature and the promotional material of South African and key international destinations indicate that in spite of its potential, the role of food in the marketing of destinations has until recently received very little attention globally, as well as in South Africa. These observations are in contrast to a study undertaken during 2000 by the WTO, where food and drink products of a destination were considered as one of the most important cultural expressions of a destination (Handszuh, 2000; Hjalager & Corigliano, 2000).

Local food enhances the appeal of a destination, and eating is an important activity for tourists once at the destination (European Commission, 1999; Selwood, 2003). Tourists seeking nature and culture are interested in

Figure 14.2 Inter-relatedness between sustainable competitiveness, food tourism and destination marketing management (du Rand, 2006)

sampling local food products and tasting authentic regional recipes. According to the Eurobarometer Survey (European Commission, 1999), more attention is being paid to the origin of the food, that is, local specialties and produced goods.

Based on these perspectives it can be argued that local food is a feature that can add value to a destination (Boyne *et al.*, 2003; Handszuh, 2000; Telfer & Wall, 1996) and contribute to the sustainable competitiveness of a destination (Crouch & Ritchie, 1999a). Knowledge regarding food tourism consumer behaviour will allow food tourism stakeholders to target and develop markets, thereby intervening in the decision-making process and persuading the consumer to purchase local food products and services (Mitchell & Hall, 2003).

Knowledge of the local, regional and national cuisine has become an interest for tourists (Chappel, 2001; Gallagher, 2001). Santich (1999) and Macdonald (2001) also report that people interested in travelling for gastronomic motivations are on the increase. In spite of these trends, gastronomy has not been considered for its real potential (Selwood, 2003), nor exploited conveniently as a tourism resource and subsequently not portraying the behaviour of the food tourism consumer. Food tourism is not purely an income-generating activity but a cultural enhancement activity contributing to the tourism experience. It therefore needs to be considered as an essential component of the marketing strategy of a destination.

Lessons from International Food Tourism Trends and Best Practices

Food tourism globally remains a form of niche tourism, based on agriculture, culture and the tourism infrastructure. It is generally linked to cultural or heritage tourism and although it forms an important component of tourism, is still in many countries very much a less promoted attraction. The attention food tourism is receiving is growing for various reasons such as changing consumer needs, environmental awareness (Ritchie & Crouch, 2000) and destination competitiveness and sustainability (Ritchie & Crouch, 2003b).

Initiatives regarding the development and implementation of food tourism range from food tourism strategies (Deneault, 2002); establishment of food tourism networks and regional development (Haas, 2002; Hall *et al.*, 2003; Murray & Haraldsdóttir, 2004); food tourism marketing activities (Hall, 2003; Handszuh, 2000); development of food events, food festivals and local product promotion (Boniface, 2003; Rusher, 2003); culinary heritage identification (Long, 2003; Ohlsson, 2000) and development of food-related activities such as gourmet cooking holidays, food and wine tours and food routes (Hall *et al.*, 2003b; Meyer-Czech, 2003; Sharples, 2003). These global and local initiatives provide sufficient reason to encourage Destination Management Organisations (DMOs) to develop similar and competitive food tourism products and activities in their own destinations.

Local Food Tourism Situation in South Africa

A situation analysis was undertaken among a representative sample of local and regional destination marketing organisations in South Africa to determine the following:

- Status of food as an attraction in destinations.
- Role of food in destination marketing strategies.
- Constraints and gaps experienced in utilising local food as attractions and as key elements of destination marketing strategies.

Detailed statistical information of the situational analysis is presented in du Rand (2006).

The status of food as an attraction in the tourism regions of South Africa

The position of food as an attraction in the tourism regions in South Africa compares with what is reported in the literature, namely that food is not primarily a key attraction, but a supportive attraction. South Africa is best known for its nature-based attractions followed by cultural and outdoor recreational activities. This provides proof that food tourism needs to be developed as a supportive attraction that can enhance visitor experiences and increase the competitiveness of the destination. The availability of agricultural resources and infrastructure provides further reason for the development of food tourism (see also Rogerson, Chapter 2 in this book).

The key components of food tourism lie strongly in the presence of local and/or regionally produced food products and speciality restaurants of the area. This is in accordance with expectations as this is the traditional way of showcasing the food of a region and offering the tourist a cultural experience. Of particular significance is the finding that routes, festivals and events are also receiving considerable attention.

The main reasons for food not being regarded as an attraction in a destination are that the region stakeholders were unaware of the tourism potential or had insufficient knowledge regarding the promotion of food. This supports the need for a food tourism marketing strategy to be developed as in countries such as Canada, Australia and the USA.

Wine and food tourism are intertwined and in South Africa, with its well-established wine routes, it is imperative that cognisance is taken of the importance and presence of wine tourism (Bruwer, 2003; Demhardt, 2003; Tassiopoulos *et al.*, 2004). Wine tourism is stronger than food tourism as a key attraction in South Africa, but does not fare that well as a supportive attraction as wine tourism is still very much localised and has recently become more of a destination experience in areas where wine is not produced.

The current role of food in regional destination marketing strategies

Regional tourism stakeholders are aware of the importance of food, as 56% of the respondents in the study reported that food is used as a marketing activity in promoting a destination. The traditional methods of marketing are utilised to a limited extent regarding food tourism, as the number of respondents utilising these methods represent only 35.7%. Together with limited use of these traditional methods of marketing, the content quality of the information regarding food is scant and uninformative. An analysis of promotional literature indicated that little food tourism information was included in the material. The focus was more on advertisements of restaurants, which supports the lack of informative food tourism information (du Rand *et al.*, 2003). Very few tour operators promote food tourism, which verifies the fact that food tourism is not a priority and lacks a focused marketing strategy. Therefore, a framework and procedure to assist DMOs in marketing and implementing food tourism in a region would be beneficial.

Constraints and gaps regarding food tourism

Nearly 60% of the respondents regarded the lack of food promotion as the key constraint or gap. The fact that there are no special food events organised in a region and that funds are a constricting factor, exacerbates the situation of food not being promoted as a special or niche tourism attraction.

Of significance are that the constraints regarding marketing (special food events, branding, media coverage and brochures) are perceived to be more important than the constraints regarding the product (routes, speciality restaurants, quality of food service and products). The lack of funds, and the fact that tourists, regardless of the marketing initiatives undertaken, will consume food may be the underlying reason for this situation. For DMOs to become more competitive and contribute to sustainable tourism the development and marketing of attractions such as food should be approached with more urgency, especially in the regions where the resources are plentiful and the infrastructure is in place.

The respondents' views on initiatives required to address the perceived gaps and constraints accentuate the fact that marketing requires attention. Promoting the use of local products and developing a marketing strategy incorporating food tourism were the most important actions proposed.

Towards a Local Food Tourism Strategy and Framework

Figure 14.3 portrays the relationship between the identified gaps regarding food tourism, and the specific actions and strategies that were recommended. Of significance are the consistent higher ratings that the marketing and promotion-related aspects, namely, media coverage, food promotion and branding, of food tourism received above that of the food product itself.

The strategies and actions recommended by stakeholders for future food tourism development accentuated the importance of developing a focused marketing strategy and framework, which includes food tourism and addresses both the product and the marketing activities. Such a framework can also facilitate the development of food tourism as an attraction in a destination. Destination competitiveness and sustainability can also be enhanced by such actions and result in the responsible utilisation of available resources and existing infrastructure.

To enable the DMOs to assess and optimise the food tourism potential of their destinations, tools such as FOODPAT, a culinary atlas and database for the tourism regions of South Africa and a Product Potential and Attractiveness Tool were developed. These tools function as decision support tools to determine the food tourism potential of a destination. The proposed Strategic Food Tourism Destination Marketing Framework and procedure for developing and implementing food tourism, as explained

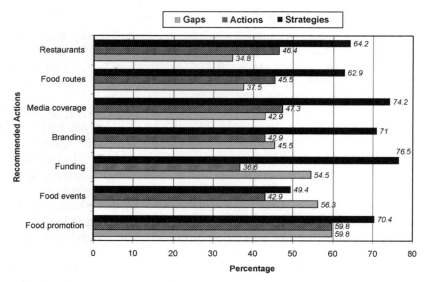

Figure 14.3 Relationship between identified gaps, actions and recommended strategies (du Rand, 2006)

in detail in du Rand (2006), is the outcome of the integration of data collected by means of the situational analysis, the evaluation of best practices, the collation of expert opinions and the information regarding food tourism acquired from FOODPAT.

Applying the Local Food Tourism Framework

The framework and its components were tested to determine its perceived relevance and viability and to verify the practical applicability of the framework in a case study situation. The assessment procedure of the framework was necessary prior to the actual implementation of such a framework, which will require a longer period of time to establish.

The framework and tools were tested with stakeholders in a leading South African regional destination, where food and wine is a key element of the overall destination experience. The case study was executed as a workshop in the Winelands area, one of the tourism regions in the Western Cape Province. The aim of the workshop was to assist destination marketers and entrepreneurs to optimise the tourism potential of local food and drink products so as to ensure sustainable competitiveness.

The 'Winelands' region was selected because it has the ability to be presented as a 'Premier-ranked Food and Wine Tourist Destination' in South Africa, that is, 'a place offering the best regarding food and wine tourism and a place the potential tourist visitor should consider first when making travel plans'. A detailed procedure followed during the implementation of the case study is outlined in du Rand (2006).

The outcomes of the case study indicated that the framework and tools provided the stakeholders with useful mechanisms to strategically and practically develop and implement food tourism as a key and/or supportive element in the positioning and marketing of regional tourism destinations. The general assessment of stakeholders was positive regarding the use of the framework and procedure for developing and implementing food tourism at regional level. Participants were also of the view that facilitation of the procedure and implementation of the framework could result in a higher level of success and could be applied in other destinations as well.

Guidelines for Sustainable Local Competitiveness

The following guidelines and recommendations can be proposed based on the information regarding the South African situation and information gained from global best practices and should form the foundation of a sustainable competitive approach for the development of food tourism.

Utilisation of food as a tourism attraction

The next guidelines and initiatives can be considered to elevate the role of food in destination marketing and to position food as a competitive and sustainable tourism attraction:

- establish appropriate marketing initiatives, partnerships and networking, supporting local products of high quality;
- improve media coverage of South African cuisine by including culinary-related information in tourism promotional material; establishment of web-based links to culinary products and experiences; inclusion of cuisine into regional tourism marketing campaigns aimed at local and global markets;
- utilise cross-marketing to enhance food and wine as a significant attraction in a destination;
- optimise current and potential markets by ensuring that standards are in place and that quality is consistently provided;
- follow 'lifestyle' positioning of food and wine tourism within the tourism strategy and support the quality of life, nature and leisure components of tourism marketing;
- adopt a 'niche' type of approach aimed at both local and international guests;
- utilise as a tool to extend the current tourism seasons;
- use as a branding tool in destinations with an attractive/unusual/ unknown cuisine;

- develop innovative signage and logos that identify food attractions in specific regions;
- improve theming, packaging and routing of food tourism by forming links with other tourism attractions and activities such as nature, sport, history and culture;
- create new experiences and provide greater impact for a destination by on-theming, for example, wine and food, food and culture, food and history and food and health;
- market and promote food tourism together with other attractions to encourage visitors to experience the wider cultural, heritage and natural characteristics of the destination, such as agri-tourism, wine tourism and ecotourism;
- linking of food routes to existing wine or historical tourism routes; and
- develop specialty restaurants to promote the special cuisine of an area.

Need for a strategic approach to develop and implement food tourism in a destination

It was recommended that a strategic approach be developed to implement food tourism at a destination. The strategic approach developed and assessed for the possible implementation indicated that the stakeholders would benefit from applying the procedures devised. It is imperative that the DMOs and other stakeholders in the various tourism regions follow a definite procedure, identify the environment, assess the resources and attractions and determine the markets for their destinations. Having proof of the food tourism potential of a destination is a crucial factor in planning for successful tourism development at a destination.

Mechanisms to develop and implement food tourism

The participant stakeholders in the case study indicated the need for facilitation with regard to the procedure of developing and implementing food tourism at a destination. Important information is required and data management is critical when determining the strategy for food tourism at a destination. Tools and procedures facilitate the process and assist the stakeholders in the strategic evaluation of the food tourism potential aspect and enable the process of planning and implementing food tourism.

Stakeholder involvement

The participant stakeholders in the case study identified the lack of stakeholder involvement and networking as one of the major short-comings. A higher level of stakeholder involvement will assist in

(1) The establishment of local and regional partnerships that can contribute to the enhancement of food tourism in the region and within South Africa. Participation of and input from many different organisations and stakeholders, namely, producers, developers and consumers of the food tourism products, are required for inter- and intra-organisational cohesion within the food and tourism industry.
(2) Identification and involvement of champions to drive the initiatives regarding food tourism and the development of a focused food tourism strategy involving all role players.
(3) Establishing community tourism initiatives and partnerships to encourage local marginalised communities to showcase their culture and heritage in a marketable manner while retaining authenticity of the products and experiences.
(4) Strengthening linkages between food tourism and other economic sectors so as to provide more local employment and income from tourism.
(5) Proper product development, supported by both the public and private sector, through mentoring and proper guidance.
(6) Improved communication and interaction among stakeholders and the media.
(7) Establishing a more focused approach for the execution of key marketing management tasks.

Stakeholders and DMOs should be encouraged to participate in promoting food tourism and to on-sell food with wine which is a better-established tourism product and attraction in many regions in South Africa.

Education, training and capacity building

The key challenge of food tourism in South Africa is the utilisation of local food as a catalyst for local pride. South Africans need to cultivate a sense of pride in their uniqueness and realise the advantages of promoting local cuisine and culinary heritage. Education, training and capacity building would assist in

(1) Developing a sense of local pride regarding indigenous cuisine and using local food to increase demand and improve availability of products and authentic experiences so that South African cuisine can be described as 'regionally based, nationally presented and globally accepted' (Hallmans, 2000: 1). The training of tourism operators and tourist guides should include information on the cuisine of the area to promote and support the food experiences offered to the visitor and increase the guides' awareness of food as a tourism product.
(2) The generation of public and private support in terms of finance, training, skills development, quality standardisation, capacity building

and mentorship is required to enable stakeholders and entrepreneurs to establish new ventures in the food tourism industry and to contribute to the delivery of consistent and quality service.

The outlined guidelines and recommendations for a way forward need to be considered by stakeholders at local and regional level prior to the development of a definite strategy for the development and implementation of food tourism on national level. Best practice frameworks, norms, standards, guidelines and ultimately a strategy are essential if South Africa is to succeed in developing as a recognised food tourism destination that is sustainable and contributes to destination competitiveness.

References

Boniface, P. (2003) *Tasting Tourism: Travelling for Food and Drink*. Hampshire: Ashgate.

Boyne, S., Hall, D. and Williams, F. (2003) Policy, support and promotion for food-related tourism initiatives: A marketing approach to regional development. *Journal of Travel and Tourism Marketing* 14 (1), 131–154.

Boyne, S., Williams, F. and Hall, D. (2002) On the trail of regional success: Tourism, food production and the Isle of Arran Taste Trail. In A. Hjalager and G. Richards (eds) *Tourism and Gastronomy* (pp. 92–131). London: Routledge.

Bruwer, J. (2003) South African wine routes: Some perspectives on the wine tourism industry's structural dimensions and wine tourism product. *Tourism Management* 24, 423–425.

Canadian Tourism Commission (2003) *How-to Guide: Develop a Culinary Tourism Product*. Ottawa: Canadian Tourism Commission.

Chappel, S. (2001) Globalization. On WWW at http://business.unisa.edu.au/cae/globlisation/abstracts.htm. Accessed 21.9.2001.

Crouch, G.I. and Ritchie, J.R.B. (1999a) Tourism, competitiveness and social prosperity. *Journal of Business Research* 44 (1), 137–152.

Crouch, G.I. and Ritchie, J.R.B. (1999b) Tourism, competitiveness social prosperity. *Journal of Business Research* 44 (1), 137–152.

Cusack, I. (2000) African cuisines: Recipes for nation building? *Journal of African Cultural Studies* 13 (2), 207–225.

Demhardt, I.J. (2003) Wine and tourism at the 'Fairest Cape': Post-apartheid trends in the Western Cape Province and Stellenbosch (South Africa). *Journal of Travel and Tourism Marketing* 14 (1), 113–130.

Deneault, M. (2002) *Acquiring a Taste for CuisineTourism* (pp. 1–15). Ottawa: Canadian Tourism Commission.

du Rand, G.E. (2006) The role of local food in destination marketing: A South African situational analysis. PhD dissertation, University of Pretoria.

du Rand, G.E., Heath, E. and Alberts, N. (2003) The role of local and regional food in destination marketing: A South African situation analysis. *Journal of Travel and Tourism Marketing* 14 (1), 97–112.

European Commission (1999) *Enhancing Tourism's Potential for Employment: European Communities*. Brussels: European Commision.

Gallagher, B. (2001) The role of food and beverage in tourism. *Keynote Address Presented at the Tourism as a Catalyst for Community Development Conference*. Pretoria, South Africa.

Haas, R. (2002) The Austrian country market: A European case study on marketing regional products and services in a cyber mall. *Journal of Business Research* 55, 637–646.

Hall, C.M. (2003) Wine food and tourism marketing: Preface. *Journal of Travel and Tourism Marketing* 14, xxiii.

Hall, C.M., Mitchell, R. and Sharples, L. (2003) In C.M. Hall, L. Sharples, R. Mitchell, N. Macionis and B. Cambourne (eds) *Food Tourism Around the World* (pp. 25–59). Amsterdam: Butterworth-Heinemann.

Hall, C.M., Sharples, L., Mitchell, R., Macionis, N. and Cambourne, B. (eds) (2003b) *Food Tourism Around the World* Oxford: Butterworth-Heinemann.

Hallmans, P. (2000) South Africa's culinary heritage. In *Execuchefs Newsletter*, February (pp. 1–6).

Handszuh, H. (2000) Local food in tourism policies. In *International Conference on Local Food and Tourism*. Larnaka, Cyprus.

Hjalager, A. (2002) A typology of gastronomy tourism. In A. Hjalager and G. Richards (eds) *Tourism and Gastronomy* (pp. 21–35). London: Routledge.

Hjalager, A. and Corigliano, M.A. (2000) Food for tourists – determinants of an image. *International Journal of Tourism Research* 2, 281–293.

Hjalager, A. and Richards, G. (2002a) In A. Hjalager and G. Richards (eds) *Tourism and Gastronomy* (pp. 224–234). London: Routledge.

Hjalager, A. and Richards, G. (eds) (2002b) *Tourism and Gastronomy*. London: Routledge.

Kaspar, C. (1986) The impact of catering and cuisine upon tourism. In *36th AIEST Congress: The Impact of Catering and Cuisine upon Tourism* Montreux, Switzerland: AIEST Publications.

Long, L. (1998) Culinary tourism. A folkloristic perspective on eating and otherness. *Southern Folklore* 55, 181–204.

Long, L. (ed.) (2003) *Culinary Tourism: Food, Eating and Otherness*. Lexington: University of Kentucky Press.

Macdonald, H.S. (2001) National tourism and cuisine forum: 'Recipes for success'. Canadian Tourism Commission.

Meyer-Czech, K. (2003) Food trails in Austria. In C.M. Hall, L. Sharples, R. Mitchell, N. Macionis and B. Cambourne (eds) *Food Tourism Around the World* (pp. 149–157). Amsterdam: Butterworth-Heinemann.

Mitchell, M.A. and Hall, C.M. (2003) Consuming tourists: Food tourism consumer behaviour. In C.M. Hall, L. Sharples, R. Mitchell, N. Macionis and B. Cambourne (eds) *Food Tourism Around the World* (pp. 60–80). Amsterdam: Butterworth-Heinemann.

Murray, I. and Haraldsdóttir, L. (2004) Developing a rural culinary tourism product: Considerations and resources for success. In *ASAC*. Quebec.

Ohlsson, T. (2000) Regional culinary heritage: A European network for regional foods. In W.T. Organisation (ed.) *Local Food and Tourism International Conference* (pp. 132–141). Cyprus: Cyprus Tourism Organization and WTO.

Quan, S. and Wang, N. (2003) Towards a structural model of the tourist experience: An illustration from food experiences in tourism. *Tourism Management* 25, 297–305.

Ritchie, J.R.B. and Crouch, G.I. (2000) The competitive destination: A sustainability perspective. *Tourism Management* 21, 1–7.

Ritchie, J.R.B. and Crouch, G.I. (2003a) *The Competitive Destination. A Sustainable Tourism Perspective* Wallingford: CABI.

Ritchie, J.R.B. and Crouch, G.I. (2003b) *The Competitive Destination. A Sustainable Tourism Perspective* Oxon: CABI.

Rusher, K. (2003) The Bluff Oyster Festival and regional economic development. In C.M. Hall, L. Sharples, R. Mitchell, N. Macionis and B. Cambourne (eds) *Food Tourism Around the World* (pp. 192–205). Amsterdam: Butterworth-Heinemann.

Santich, B. (1999) Location, location, location. *The Age*. On WWW at http://www.theage.com.au/daily/990622/food/food.1.html.

Selwood, J. (2003) The lure of food: Food as an attraction in destination marketing in Manitoba, Canada. In C.M. Hall, L. Sharples, R. Mitchell, N. Macionis and B. Cambourne (eds) *Food Tourism Around the World* (pp. 178–191). Amsterdam: Butterworth-Heinemann.

Sharples, L. (2003) The world of cookery-school holidays. In C.M. Hall, L. Sharples, R. Mitchell, N. Macionis and B. Cambourne (eds) *Food Tourism Around the World* (pp. 178–191). Amsterdam: Butterworth-Heinemann.

Tassiopoulos, D., Nuntsy, N. and Haydem, N. (2004) Wine tourists in South Africa: A demographic and psychographic study. *Journal of Wine Research* 15, 51–63.

Telfer, D. and Wall, G. (1996) Linkages between tourism and food production. *Annals of Tourism Research* 23, 635–653.

Wagner, H.A. (2001) Marrying food and travel ... culinary tourism. In *Canada's Food News*. Foodservice Insights.

Wolf, E. (2002) *Culinary Tourism: A Tasty Economic Proposition*. Portland, Oregon: International Culinary Tourism Task Force.

World Tourism Organisation (WTO) (1999) *Guide for Local Authorities on Sustainable Tourism Development* Madrid: World Tourism Organisation.

Yoon, Y. (2002) Development of a structural model for tourism destination competitiveness from stakeholders perspectives. PhD dissertation, Virginia Polytechnic.

Chapter 15

Conclusions and Critical Issues in Tourism and Sustainability in Southern Africa

JARKKO SAARINEN

Introduction

It is widely acknowledged that the growth and present scale of tourism has led to a range of evident problems. These include not only environmental, social, cultural and economic issues in destination areas and communities but also changes in tourists' home regions, routes and transit regions, including impacts on global environmental change. Such problems and threats have created an urgent need for more sustainable principles and practices in tourism. In addition, the industry has become more aware of the nature of its impacts and its role in the changing environment, and how these changes may affect the industry's future potential. Especially, the issue of climate change has been highlighted (see Gössling & Hall, 2005). In spite of these positive changes, it is clear that the current modes and growth of tourism cannot be seen possible or acceptable without some interference in the industry's practices in the future (Ashworth & Dietvorst, 1995; Mowforth & Munt, 1998).

Presently, the idea of sustainability affects and guides – or at least it should guide – the development and management processes of the industry at all the levels and scales of the tourism system. Sustainability should be built up both on the supply and on the demand side of contemporary tourism. It already has a great influence on the policies and politics of tourism development, as indicated in most of the case studies in this book. Sustainability and related indicators also provide elements, by which local people, tourists and policy-makers may evaluate and appreciate destinations and different types of tourism (see Hall, 2000; Holden, 2007; Sharpley, 2000). However, there are still many interpretational and practical problems inherent in the concept of sustainable tourism, and especially in its relation to sustainable development. One of the characterising problems is tied to the multi-level and holistic nature of sustainable development. The division of sustainable tourism into ecological, socio-cultural and economic dimensions is an analytical one, which helps us to understand the multivalent nature of sustainability, but the different dimensions should finally be understood and defined as interactive processes, or as an entity made up of all three components (see Martin & Uysal, 1990; Wolfe, 1983).

Another challenge created by the multi-level holistic perspective is connected with the spatial and functional scales (Sharpley, 2000). The focus of sustainability in tourism is still set mainly on the existing tourist activities in the destination areas (Saarinen, 2006). However, in spite of the problems, it is also important to look beyond the present impacts and effects at the destinations in order to create tourism development that is truly sustainable (Gössling & Hall, 2005; Wall, 1996). As indicated by Bramwell and Lane (2008: 1), the new emphasis on sustainable tourism should be more global in scale (see Holden, 2003, 2007). Naturally, the 'new sustainable tourism' with a more integrative and holistic approach is not an easy one to implement in practice (see Inskeep, 1991) because it is almost impossible to consider the whole tourism system, all the impacts and their relations to non-touristic processes, all the spatial and temporal scales involved in the production of tourism. Still, other scales than the local destination level should be recognised, and possible successful implementations and practices of sustainable tourism should be placed in the wider context of tourism and local social, economic and ecological systems.

Critical Aspects of Tourism and Sustainability

The reason why the focus of sustainable tourism development has been somewhat narrow may lie in the nature of the industry and tourism itself. In general, tourism refers to the movement of people and their stay in space (destinations and transit areas). Although the spaces of tourism are nowadays increasingly being used and consumed virtually (see Bristow, 1999; Buhalis, 2000; Dewailly, 1999; Ravenscroft, 1999), tourism as an idea that still mainly requires factual movement of people, related goods, capital, ideas and values with physical and socio-cultural impacts, for instance. Tourism is increasingly a part of an expanding field of mobility, which is seen as a new kind of paradigm in tourism studies and social sciences (see Hall, 2005; Hannam, 2008; Urry & Sheller, 2006). In that respect, tourism is not simply a matter of movement and staying at a tourist destination, but it also involves a multitude of flows and experiences of the environment and of being a tourist (or a host) (Gordon & Goodall, 2000). One important aspect of being a tourist or a host is consumption; tourism is an industry and a sector of the economy, which creates touristic commodities and spaces of consumption (see Sack, 1992). Consumption characterises tourism phenomena including the tourists, the hosts and their relationship. In a way, the whole tourist destination, such as Sun City and Kruger National Park in South Africa, Swakopmund and Etosha National Park in Namibia, Victoria Falls in Zambia and Zimbabwe or Okavango Delta and Kalahari (Kgalagadi) in Botswana, can be seen as a commodity in tourism that is framed, developed and produced for non-local people to see and consume.

Indeed, this aspect of non-locality is a key issue that has previously jeopardised and still challenges the industry's future sustainability. Most of the classical and technical definitions of tourism recognise it as a short-term leisure and recreation activity for people who are away from their normal environment of residence and work with the intention of returning to their home community within a few days, weeks or months (Burkhart & Medlik, 1981: 42). However, presently not only tourists are mobile, but also the industry, related activities and impacts. Particularly in peripheries, tourism is often developed and controlled mainly by non-local actors, investors and capital, and/or it may represent non-local change, value systems and ideologies for the local communities (see Greenwood, 1989; Urry, 1990; Waitt, 1999). As Getz (1999: 24) has stated, a 'tourist space is an area dominated by tourist activities or one that is organised for meeting the needs of visitors.' Thus, from the industry's perspective, the needs of (non-local) visitors are the leading guidelines, which may risk the sustainability of 'sustainable tourism' or 'community-based tourism' in practice, if local needs, expectations and values systems are much different than the visitors' and industry's.

The evolution of sustainable and other alternative forms of tourism is related to our growing knowledge pertaining to negative impacts of tourism, the perceived potential of tourism to contribute to sustainable development and, more recently, to poverty reduction and to societal shifts emphasising 'greener' modes of production and consumption (Saarinen, 2006). Mowforth and Munt (1998) recognise these elements, but more cynically, they see the discussions and attempts at developing sustainable tourism 'as the "natural" continuation of the process of colonialism and control', merely presented in a new guise (see also Pattullo, 1996). Especially in the Developing World and their peripheries, tourism and travel have always been more or less externally managed and guided colonialistic acts (Britton, 1991), and according to Mowforth and Munt sustainable tourism is no exception to this. Therefore they describe sustainable tourism as the mainstream tourism industry's attempt to invent a new legitimatisation for itself: 'sustainable and rational use' of the environment, including the preservation of nature as an amenity for the already advantaged (Mowforth & Munt, 1998: 96). On the other hand, it is fair to state that tourism is often also relatively beneficial for the development of its destination regions by providing job opportunities, income and possibilities for entrepreneurship. However, that alone does not make it sustainable.

This contradiction between the different views is evident in the southern African context and well demonstrated in the book with the cases related to the Okavango Delta and Transfrontier parks (see Mbaiwa & Darkoh, Chapter 12 and Ramutsindela, Chapter 10 in this book), for example. Based on these cases, it is obvious that tourism can clearly bring benefits to people and open avenues for nature conservation processes, but locally it can also create negative impacts which may jeopardise the long-term sustainability in tourism development. In the end, sustainable tourism is an inseparable part of the industry; and to exist, the industry and its presence are to be taken as granted, that is, tourism and the industry are the natural starting points and conditions for the conceptual basis for sustainable tourism. As a result of this, the principles of sustainable development cannot be easily transposed onto (sustainable) tourism as a specific economic and social activity (see Butler, 1993; Wall, 1993). Due to these contradictions, sustainable tourism is often seen as a marketing strategy, for example, that is used as 'green' labelling of the business (see Wheeller, 1993). Indeed, there are tens, if not hundreds, of different certificates that are referring to sustainability, environmentally responsible tourism or ecotourism, but their background criteria may differ greatly and there may not be factual or monitored systems looking at the impacts and benefits of the specific operations. Without having transparent and generally acknowledged standards it may be very difficult to separate a green from a greenwashed product, and there are still very few such widely acceptable standards for the industry (Box 15.1).

Box 15.1   Global Sustainable Tourism Criteria

United Nations Environment Programme (UNEP), United Nations World Tourism Organisation (UNWTO), and United Nations Foundation and Rainforest Alliance announced Global Sustainable Tourism Criteria (GSTC) at the World Conservation Congress in 2008. The criteria were developed by the Partnership for GSTC, which is a coalition of 27 organisations including tourism leaders from the private, public and non-profit sectors. The coalition aims to increase an understanding of sustainable tourism practices and the adoption of universal sustainable tourism principles. To establish a Global Criteria canon, the partnership coalition reviewed over 60 certifications and criteria globally and analysed more than 4500 existing criteria.

The established GSTC focus on four main areas (A–D) with specific sub-aims:

Demonstrate effective sustainable management

(1) The company has implemented a long-term sustainability management system that is suitable to its reality and scale, and that considers environmental, socio-cultural, quality, health and safety issues.
(2) The company is in compliance with all relevant international or local legislation and regulations (including, among others, health, safety, labour and environmental aspects).
(3) All personnel receive periodic training regarding their role in the management of environmental, socio-cultural, health and safety practices.
(4) Customer satisfaction is measured and corrective action taken where appropriate.
(5) Promotional materials are accurate and complete and do not promise more than what can be delivered by the business.
(6) Design and construction of buildings and infrastructure.
(7) Information about, and interpretation of, the natural surroundings, local culture and cultural heritage is provided to customers, as well as explaining appropriate behaviour while visiting natural areas, living cultures and cultural heritage sites.

Maximise social and economic benefits to the local community and minimise negative impacts

(1) The company actively supports initiatives for social and infrastructure community development including, among others, education, health and sanitation.
(2) Local residents are employed, including in management positions. Training is offered as necessary.
(3) Local and fair-trade services and goods are purchased by the business, where available.
(4) The company offers the means for local small entrepreneurs to develop and sell sustainable products that are based on the area's nature, history and culture (including food and drink, crafts, performance arts, agricultural products, etc.).
(5) A code of conduct for activities in indigenous and local communities has been developed, with the consent of, and in collaboration with, the community.

(6) The company has implemented a policy against commercial exploitation, particularly of children and adolescents, including sexual exploitation.

(7) The company is equitable in hiring women and local minorities, including in management positions, while restraining child labour.

(8) The international or national legal protection of employees is respected, and employees are paid a living wage.

(9) The activities of the company do not jeopardise the provision of basic services, such as water, energy or sanitation, to neighbouring communities.

Maximise benefits to cultural heritage and minimise negative impacts

(1) The company follows established guidelines or a code of behaviour for visits to culturally or historically sensitive sites, to minimise visitor impact and maximise enjoyment.

(2) Historical and archaeological artefacts are not sold, traded or displayed, except as permitted by law.

(3) The business contributes to the protection of local historical, archaeological, culturally and spiritually important properties and sites, and does not impede access to them by local residents.

(4) The business uses elements of local art, architecture or cultural heritage in its operations, design, decoration, food or shops; while respecting the intellectual property rights of local communities.

Maximise benefits to the environment and minimise negative impacts

(1) Conserving resources.

(2) Reducing pollution.

(3) Conserving biodiversity, ecosystems and landscapes.

The criteria intend to set a minimum standard that any tourism business should aim to reach in order to protect and sustain natural and cultural resources while ensuring tourism meets its (full) potential as a tool for poverty alleviation. In this respect, the criteria aim to be part of the response of the tourism industry to UN Millennium Development Goals (MDGs).

More information from http//www.SustainableTourismCriteria.org.

Setting the criteria, that is, limits to growth is an urgent issue in southern Africa and for the sustainability of the tourism development in the region. For example, in relation to the Okavango Delta and the recent release of the new management plan of the area Ian Michler (2008: 26) highlights 'the need to establish strict criteria that detect when "limits of use" are being exceeded. Without these, the region is doomed to become another over-utilised game park that is used to meet the financial demands of stakeholders.' Indeed, there may be a factual threat of exploitation if all the non-conservationist stakeholders are aimed to be served and satisfied in development (see Mabunda & Wilson, Chapter 7 in this book). Still, sustainability is more than setting the limits to growth based on one element of sustainability, and on a principal level there is no difference whether the one preferred is based on ecology, economic or social matters: in the context of sustainable development that is insufficient.

Repositioning Tourism

The tourism industry has a tremendous capacity for generating growth and wealth in destination areas and communities in southern Africa. Sustainable tourism development is regarded as a possible mechanism to break the cycles of poverty and environmental degradation and to contribute to the achievement of the UN Millennium Project's Development Goals (MDGs) in the continent (Box 15.2). Especially the goals related to poverty alleviation, environmental conservation, gender equality and global partnership for development have been emphasised. However, according to Telfer and Sharpley (2007) it is still unclear to what extent tourism can be used effectively as a medium of addressing these extreme challenges in development. Similarly, Zhou and Richie (2007: 120) state that the relation 'between tourism and poverty alleviation largely remains *terra incognita* among tourism academics'. Still, the matter that is clear and needs to be mapped is the need to re-position tourism in sustainable development discourses.

Box 15.2   United Nations' Millennium Project's Goals

The UN Millennium Project has eight MDGs that are targeted to be met with the slogan 'end poverty 2015' (www.un.org/milleniumgoals):

Goal 1: Eradicate extreme poverty and hunger.
Goal 2: Achieve universal primary education.
Goal 3: Promote gender equality and empower women.
Goal 4: Reduce child mortality.
Goal 5: Improve maternal health.

Goal 6:   Combat HIV/AIDS, malaria and other diseases.
Goal 7:   Ensure environmental sustainability.
Goal 8:   Develop a global partnership for development.

Sustainable development and MDGs, for example, are primarily linked with the needs of people – not a certain industry – and the use of natural and cultural resources in a way that will also safeguard human needs in the future (Redcliffe, 1987; Spangenberg, 2005; WCED, 1987). The needs of people and those of the tourism industry are not necessarily conflicting, but as noted by Butler (1999), tourism may not always be the most favourable or wisest use of natural or cultural amenities and resources in specific locations in the long term. Thus, 'sustainable tourism' may in practice be an unsustainable and unequal process for the original local communities or natural habitats (Bianchi, 2004; Cohen, 2002), especially if local people have no right to say 'no' to tourism development. This element of control over the activities of tourism is highly related in the context of land use, ownership and related reforms in southern Africa as identified in the chapters by Becker, Govender-van Wyk and Wilson, Hall, Mbaiwa and Darkod and Ramutsindela.

The important question is on what conditions sustainable tourism could represent sustainable development locally and also in a local–global nexus. Are the present local solutions to global challenges such as the global climate change and the achievement of MDGs enough action, and do they represent all that tourism can do in respect to sustainable development? There are no single or simple answers to these questions. Basically, to practice truly sustainable tourism based on the idea of sustainable development, the position of the industry should be re-evaluated and re-located in the current development discourses and actions in southern Africa. On a local scale, the community-based tradition of sustainability aims to promote this decentralisation, with an emphasis on empowerment and strong integration into surrounding social and spatial structures and their goals. The community-based tradition is connected to the idea that tourism can contribute to a better social, economic and environmental future on a local scale by stressing the needs of local people. From the sustainable development perspective, the sustainable use of resources, the environment and the well-being of communities are goals to which sustainable tourism could and should contribute – if the industry's role is also seen to be beneficial to that process by groups other than the industry itself. Without that emphasis, the current mode and goals of sustainable tourism 'do not necessarily contribute to those of sustainable development' (Hunter, 1997: 851).

Tourism and community-based resource management literature include many different participation channels in planning and development processes (see Hall, 2000; Jones & Murphree, 2004). One of the main principles in participation processes is related to the need to consult local people, and that their ideas, preferences and views should somehow be accepted and valued in tourism planning and legislation. For a private and often non-locally owned and/or guided tourism business this may be a relatively challenging responsibility, and there are also other difficulties that are related not only to 'participation in practice', but also to the 'idea of community'. Local communities are not homogeneous; their inner hierarchies and power structures change in time and place. When considering the cases exemplified in the book, relating to Transfrontier Parks, the Okavango Delta, Namaqualand or the Isimangaliso Wetland Park, the basic question that can be raised is: who actually represents local communities in planning processes and who are the main stakeholders they are usually listening to? And if there are conflicting arguments among the local people and stakeholders, as often is the case, how to weight and integrate the different thoughts and ideas into the planning and development processes in principle and in practice. It is important to note that when the local community consists of different groups with different preferences regarding tourism and its limits of growth (Lew, 1989), these different groups are not necessarily equally represented or involved in participatory processes (Akama, 1996; Getz & Timor, 2005; Kieti & Akama, 2005). This situation makes the participatory approach a challenging issue, but one that is crucial to the study in the context of sustainable tourism developments in southern Africa.

Related to participation, Tosun (2000) has identified three major challenges in community participation that are operational, structural and cultural by their nature. Operational limits of participation refer to problems in communication and information, sharing and lack of coordination between different policy levels and to the actors involved. Structural limits include power issues and organisational barriers, whereas cultural limits of participation indicate cultural and knowledge differences between tourism development actors and local people (see Ramutsindela, Chapter 10 in this book). Cultural limits do neither support the tourism industry to integrate local people into planning and development nor encourage local communities to participate (Saarinen & Niskala, 2009). Whether there are cultural limits or other barriers in participatory processes or not, local communities should be addressed in tourism aiming at sustainable development. However, at the same time – and from the very same perspective of sustainable development – it is essential to acknowledge that local people, communities and cultures do neither have automatic granted privileges over ethical or sustainable aspects of tourism, development or related ecological issues, nor do they necessarily have intrinsic knowledge on the impacts of tourism development and the scale of such impacts on the environment. This double-sided nature of sustainability, emphasising the participation needs of local people on the one hand, while stressing the general principles of holism, equity and futurity in sustainability on the other hand, makes it a very difficult, but much needed, process to achieve in factual planning and development objectives as indicated in the case studies (see Moswete *et al.*, Chapter 11 in this book).

Indeed, these challenges of sustainability and tourism development are well demonstrated in the book. Although the initial chapters of Part I overviewed and contextualised the critical issues of development, policies, sustainability and gender in tourism, Part II and III set the focus on specific cases, places and people. In Part II, the emphasis is laid on local policies of sustainability in tourism. All the related chapters clearly indicate the urgent need for well-formulated tourism policies by the southern African governments and regional authorities with local and global level involvement and private–public sector collaboration. In the Namibian context, both Becker and Scholz underlined the role of governance in sustainable tourism development. Scholz (see Chapter 9, this volume) focused on the specific issue of how the Namibian government has approached the challenge underlying the objective of conserving resources for the benefit of communal areas and especially the rural poor. In a complementing approach, Becker (see Chapter 6, this volume) discussed the relationship between sustainable tourism, conservation and the necessity to integrate local communities into regional planning and the spatial transformation process of the country.

Further, Atlhopneng and Mulale (see Chapter 8, this volume) in Botswana and Ramutsindela (see Chapter 10, this volume) in the southern Africa Transfrontier Conservation Area creation context concluded the need to empower local communities and recognise the rights of local people to benefit from tourism and related activities. Both cases stressed that the governmental bodies should aim to activate the effective participation of local communities to achieve sustainable development in tourism. In this respect, the legal and conflict management frameworks, adopted in addressing the impacts of tourism and nature conservation, should take into account not only present economic, ecological and land-use forms but additionally historical, anthropological, sociological and ethnological perspectives. The same issues were approached from a different side by Mabunda and Wilson (see Chapter 7, this volume) who discussed the question of whether protected areas could be commercialised to generate revenue more effectively and achieve economic viability. This was done by analysing the customer perceptions towards the outsourcing plans for tourist accommodation in the rest camps of the Kruger National Park. Based on the case study, it is clear that the majority of tourists did not want accommodation to be outsourced, and they feared not only the costs but also a loss of the 'Kruger Culture' through commercialisation ('Disneyfication') of the Park. Thus, the places of tourism are symbolic for the local communities and can also be meaningful to tourists.

Although the previous chapters discussed some of the social aspects of sustainable tourism and policy issues on a local scale, Part III focused more clearly on the connections of tourism, local communities and changing environments. Moswete *et al.* (see Chapter 11, this volume) discussed the challenges of community-based tourism in peripheral areas of Botswana where local communities are dependent on government welfare programmes. This mode of dependency seems to be transmitted also to tourism activities that are guided and managed mainly from outside. Based on the literature, this seems to be a rather common feature of tourism in peripheral areas of southern Africa (see Mbaiwa, 2005; Novelli & Gebhardt, 2007; van Veuren, 2001). However, while the benefits of community-based tourism initiatives have not been fully realised in the case study areas, the potential remains high. According to Moswete *et al.*, the main challenge is to empower communities through tourism and business education, and to facilitate collaboration and trust between different stakeholders and social groups in the communities. This clearly refers to the idea of social capital and the need to empower local communities by emphasising the protection and creation of such capital (see Jones, 2005).

Similarly, Mbaiwa and Darkoh (see Chapter 12, this volume) underlined the potential and positive contribution of tourism in the Okavango Delta area and discussed in detail the negative elements of tourism from the community perspective. Based on their analysis, it seems that the negative socio-cultural and economic impacts of tourism development are indicating that the evolution of the tourism industry in the Delta has not yet given enough consideration to issues of sustainability in development. According to them there are three principal-level concerns: social equity, economic efficiency and ecological sustainability. In addition, there are other negative impacts created by tourism development such as the emergence of enclave tourism, displacement of traditional communities, demonstration effect, prostitution and poor living and working conditions of safari workers.

Govender-van Wyk and Wilson (see Chapter 13, this volume) discussed the relation of tourism, communities and resources in the transforming context of land reform and the specific land redistribution programme called the commonage sub-programme. Their case study clearly demonstrates that tourism ventures could potentially provide sustainable livelihoods for rural people if managed accordingly, and if supported by policy structures. Finally, the last case study by du Rand and Heath (see Chapter 14, this volume) aimed to demonstrate the value of locally produced and processed food. Based on international literature, they argued that local food holds great potential in southern Africa to contribute to sustainability in tourism by enhancing the local and regional tourism resource base, adding value to the authenticity of the destination and strengthening the local economy and its inter-linkages.

The common perspective in the chapters includes the need for effective and also locally acceptable tourism policies of sustainable tourism, development of joint ventures and government encouragement in local development and participation in tourism. In addition to a general emphasis on local involvement, the empowerment of different ethnic groups and women was stressed in the chapters. Further, the local control and shared responsibilities in the local–global nexus in tourism development were highlighted together with a need to improve the legal policy frames, influencing and guiding tourism development and its local connections including the distribution of benefits and power in regional and local development.

Future Agenda for Sustainable Tourism Studies

Locally, tourism and its development often mirror processes of globalisation. In the context of globalisation, the relations between sustainable development, tourism and localities are complex. It is evident that the discourses regarding the appropriate limits of growth in tourism in southern Africa and in the specific destinations in the region are not only an issue for local-scale discussions but also for global ones. This is very evident in relation to the present climate change discussions, which, for example, question the current position of international air transportation outside the Kyoto Protocol (see Gössling & Hall, 2005). Indeed, many of the global-scale processes are increasingly coming to determine local realities and practices but, as Teo (2002) noted, not all such processes that allude to globalisation are solely driven from 'out there', as local people and actors also contribute to them and their local outcomes. This may not be entirely applicable in CO_2 taxation policy contexts. It is, however, evident in tourism and sustainable development discussions in which different social groups can define and contest the appropriate goals, methods and levels for the utilisation of natural and cultural resources, and different conceptualisations of nature and culture through negotiation (see Macnaghten & Urry, 1998: 29–31; Wall, 1996). In this kind of 'globalisation from below' in the local–global nexus, the intensity of the elements of social capital (networks, norms and trust) in communities affects the possibilities of the local people to control or influence the place-specific outcomes of globalisation and the limits of growth in tourism (Jones, 2005; Teo & Lim, 2003).

Therefore, it is important to realise that the determining of sustainability and the limits of growth in tourism is not a one-way street in the global–local nexus; these limits cannot be established and based solely on local or global perspectives. Sustainability is a matter of both local and global responsibilities. In this respect, globalisation challenges many aspects of the sustainability introduced into tourism (see Bramwell & Lane, 2008) and urges different political and economic actors to place a stronger emphasis on human relations and ethics. Instead of local-scale, tourism-centric approaches, tourism as an activity needs to be decentralised in discourses and practical application, referring to sustainable development. Also from a local community perspective, the need to decentralise tourism in sustainable tourism development discussions is evident; tourism should be placed in the local and regional socio-economic context where well-being, sustainable use of resources and the environment are the goals, where tourism could and should contribute in synergy with other locally based or operating economies and functions. By locating the place, its social and physical environment and needs in the centre, the role of tourism as a potential tool, not as a goal, for sustainable development becomes understood.

The challenges embedded in the idea and practices of sustainable tourism and the issue of globalisation make the present situation an important time for the study of tourism development, destinations and their transformations in southern African context. The UNWTO, for example, has identified both poverty reduction and climate change as major challenges to the tourism industry. These two issues are crucial for the future of tourism and sustainable tourism in southern Africa, but where the potential of tourism should be further studied and developed. Thus, there is an urgent need for research into models of benefit sharing and the politics and political economy of tourism (Bianchi, 2004; Hall, 1994) in order to define desired goals and conditions, its resources and limits and how power relations and decision-making processes are established and perceived. In southern Africa, these needs relate to the urgent issues of community involvement, participation and control, and the consequences of political and economic processes of conservation, land reform and distribution programmes as indicated by Athlopheng and Mulale (Chapter 8), Govender-van Wyk and Wilson (Chapter 13), Hall (Chapter 3) and Ramutsindela (Chapter 10) in this book. This also calls for well conceived but realistic tourism policy objectives and partnership models to be developed in the region as stated by Dieke (2000: 312).

Most tourism activities in southern Africa are directly dependent on climate and weather. This is especially true for wildlife tourism, where activities are produced in the natural environments, and where natural resources and conditions play an important role (see Atlhopheng & Mulale, Chapter 8 in this book). Tourism is also a major contributor to that change and is responsible for changing landscapes, land-use patterns and lifestyles, and for emitting greenhouse gases through *inter alia* transportation and tourist activities. Subsequently, future studies should include not only ecological and physical but also global social, cultural, economic and political changes and their interrelationships with tourist activities, and how the tourism industry is contributing to the global changes in southern Africa. Greater responsibility also requires supporting policy structures, planning practices and capacity building.

In spite of the policy structures and planning, tourism and its development will always have an impact on destinations and the whole 'tourism system'. In the context of sustainable development on a local scale, the key issue in the use of cultural and natural resources ('capital') is the role of local people and their cultures and ability to influence and even control development and its local outcomes. As indicated by Manwa (Chapter 4, in this book), this includes the gender aspect and the empowerment of women in tourism (see Scheyvens, 2002). Again, it is important to remember that sustainability or sustainable development does not primarily deal in a social (or any other) sense with certain economies and industries such as tourism, but with people and communities having their own histories, traditions and value systems: tourism (or no tourism!) may represent an option and processes, by which communities and indigenous cultures may achieve sustainable development on a local, and hopefully, a larger scale (see Govender-van Wyk & Wilson, Chapter 13; Mbaiwa & Darkoh, Chapter 12 in this book). In this respect, the idea of development and related power connections should be critically evaluated in tourism studies (Church & Coles, 2007; Telfer & Sharpley, 2007).

In southern Africa, a major issue related to the connections between development and power is linked to the community involvement in tourism products and the possible related policies (see Bramwell & Meyer, 2007). In addition to participation issues in planning, it is also a question of the nature of products. Much of the tourism in the region is based on wildlife, especially in rural and peripheral areas. National parks, game reserves and other wildlife management and communal areas provide opportunities for tourists to view wildlife, for local people to get employment or for communities to lease their land to tourism providers. This can be beneficial for the conservation, local employment and regional development, but there may be higher benefits to achieve if social and cultural aspects of localities were employed in tourism independently or complementary to the main attractions based on nature and wildlife (see Manwa, 2007). In this respect, one fruitful perspective on the sustainable use of local culture, people, history and traditions in the southern African context would be a development of *heritage tourism*. Heritage refers to places, objects or ideas that are culturally and socially valued and that have been passed from one generation to the next (Prentice, 1994; Timothy & Boyd, 2003), and there is already demand for such tourism developments in the region. For example, by 1996 almost a third of international tourists had visited a cultural village during their stay in South Africa (van Veuren, 2004: 140).

South Africa Tourism (SAT) has identified cultural tourism as one of the country's key development areas (see Ivanovic, 2008), and recently many countries in the region, including Botswana, Namibia and Swaziland, have developed or are developing cultural tourism and villages in their peripheries to activate tourism and local development connections (see Saarinen, 2007). By utilising culture and traditions in tourism – instead of mainly wildlife and natural landscapes – it would perhaps be easier to build more equal and mutually beneficial partnerships between the industry and local communities. That would also potentially empower women (see Manwa, Chapter 4 in this book) and local people in general to be actively involved in the industry and tourism, instead of working mainly in maintenance tasks in conservation areas and lodging establishments, or receiving relatively minor benefits from land leasing contracts (see Novelli & Gebhardt, 2007). At the same time, cultural tourism would probably produce locally based employment outside the tourism sector more effectively for a wide range of traditional economies; for example, in the production of cultural artefacts, crafts and art (see Manwa, 2007; Turco, 1999).

However, the use of people and indigenous cultures in tourism is not without risks (see Butler & Hinch, 2007; Greenwood, 1989). In addition to the level and nature of involvement and participation, a key issue is how communities and people are depicted and used in tourism products and practices (see Mowforth & Munt, 1998; Saarinen & Niskala, 2009). This raises the issues of representation, authenticity and the sustainability of their uses in tourism. In the context of cultural villages in South Africa, van Veuren (2004: 150) concludes that it is evident that they are often built on an understanding of what western tourists want, and how best to provide this image to the non-local customers (Saarinen, 2007; van Veuren, 2001). Therefore, cultural products based on localities may actually be constructs of western culture; as stated by Wels (2004: 90) outsiders often 'want to see Africans and the African landscape in the way they are taught to see them in their formative years of image moulding', based on processes from the colonial period. To be sustainable, the past should be 'performed' or 'produced' on the basis of local knowledge, values and heritage with people, who continue to interact and have 'assimilated' that culture in both the present and the past (Lowenthal, 1985: 410–412) – and not only the historically constructed gaze of travellers and tourists (see Howes, 1996; Urry, 1990); as is often the case with indigenous people and cultural minority groups (see Del Casino & Hanna, 2000; Greenwood, 1989; Waitt, 1999).

To sum up, in respect of the future agenda for sustainable tourism in southern Africa, the mission and value of academic studies concerning the limits of growth are in evaluating and providing perspectives on the sustainable and ethical use of nature and culture in both global and local development processes. To provide such a base for tourism development, there is a need to appreciate and accept values, principles and criteria over present-day issues, affecting specific existing or potential destinations and their economic activities. Since the ethical element in sustainability is built upon both theory and practice, and on both, local and global scales of supply and demand, the industry will eventually have to change and redirect its position in planning and development discourses, provided that it really aims to promote sustainable development and truly become an activity that 'meets the needs of present tourists and host regions while protecting and enhancing opportunities for the future', as outlined by the (United Nations) World Tourism Organisation (WTO, 1993: 7).

References

Akama, J. (1996) Western environmental values and nature-based tourism in Kenya. *Tourism Management* 17, 567–574.

Ashworth, G.J. and Dietvorst, G.J. (eds) (1995) Conclusion: Challenge and policy response. In *Tourism and Spatial Transformations* (pp. 329–339). Oxon: CAB International.

Bianchi, R. (2004) Tourism restructuring and the politics of sustainability: A critical view from the European Periphery (The Canary Islands). *Journal of Sustainable Tourism* 12 (4), 495–529.

Bramwell, B. and Lane, B. (2008) Editorial: Priorities in sustainable tourism research. *Journal of Sustainable Tourism* 16 (1), 1–4.

Bramwell, B. and Meyer, D. (2007) Power and tourism policy relations in transition. *Annals of Tourism Research* 34 (3), 766–788.

Britton, S. (1991) Tourism, capital and place: Towards a critical geography of tourism. *Environment and Planning D: Society and Place* 9, 451–478.

Bristow, R.S. (1999) Commentary: Virtual tourism – the ultimate ecotourism? *Tourism Geographies* 1 (2), 219–225.

Buhalis, D. (2000) Tourism and cyperspace. *Annals of Tourism Research* 28 (1), 232–235.

Burkhart, A. and Medlik, S. (1981) *Tourism: Past, Present and Future*. London: Butterworth-Heinemann.

Butler, R. (1993) Tourism – An evolutionary perspective. In J.G. Nelson, R. Butler and G. Wall (eds) *Tourism and Sustainable Development: Monitoring, Planning, Managing* (pp. 26–43). Waterloo: University of Waterloo.

Butler, R. (1999) Sustainable tourism: a State-of-the-art review. *Tourism Geographies* 1 (1), 7–25.

Butler, R. and Hinch, T. (eds) (2007) *Tourism and Indigenous People: Issues and Implications*. Oxford: Butterworth-Heinemann.

Church, A. and Coles, T. (eds) (2007) *Tourism, Power and Space*. London: Routledge.

Cohen, E. (2002) Authenticity, equity and sustainability in tourism. *Journal of Sustainable Tourism* 10 (2), 267–276.

Del Casino, V.J. Jr. and Hanna, S.P. (2000) Representation and identity in tourism map spaces. *Progress in Human Geography* 24 (1), 23–46.

Dewailly, J-M. (1999) Sustainable tourist space: From reality to virtual reality. *Tourism Geographies* 1 (1), 41–55.

Dieke, P. (2000) Tourism and Africa's long-term development dynamics. In P. Dieke (ed.) *The Political Economy of Tourism Development in Africa* (pp. 301–312). New York: Cognizant.

Getz, D. (1999) Resort-centred tours and development of the rural hinterland: The case of Cairns and the Atherton Tablelands. *The Journal of Tourism Studies* 10 (2), 23–34.

Getz, D. and Timor, S. (2005) Stakeholder involvement in sustainable tourism: Balancing the voices. In W. Theobald (ed.) *Global Tourism* (pp. 230–247). Burlington: Elsevier.

Gordon, I. and Goodall, B. (2000) Localities and tourism. *Tourism Geographies* 2 (3), 290–311.

Greenwood, D.D. (1989) Culture by the pound: An anthropological perspective on tourism as cultural commoditisation. In V.L. Smith (ed.) *Host and Guest: The Antropology of Tourism* (pp. 171–185). Philadelphia: University of Pennsylvania Press.

Gössling, S. and Hall, C.M. (2005) An introduction to tourism and global environmental change. In S. Gössling and C.M. Hall (eds) *Tourism and Global Environmental Change* (pp. 1–33). London: Routledge

Hall, C.M. (1994) *Tourism and Politics: Policy, Power and Place*. Chichester: Wiley.

Hall, C.M. (2000) *Tourism Planning*. Harlow: Prentice-Hall.

Hall, C.M. (2005) *Tourism: Rethinking the Social Science of Mobility*. Harlow: Prentice-Hall.

Hannam, K. (2008) Tourism geographies, tourism studies and the turn towards mobilities. *Geocompass* 2 (1), 127–139.

Holden, A. (2003) In need of new environmental ethics for tourism. *Annals of Tourism Research* 30 (1), 94–108.

Holden, A. (2007) *Environment and Tourism*. London: Routledge.

Howes, D. (1996) Cultural appropriation and resistance in the American west: Decommodifying 'indianness'. In D. Howes (ed.) *Cross-Cultural Consumption: Global Markets, Local Realities* (pp. 38–160). London: Routledge.

Hunter, C. (1997) Sustainable tourism as an adaptive paradigm. *Annals of Tourism Research* 24, 850–867.

Inskeep, E. (1991) *Tourism Planning: An Integrated and Sustainable Development Approach*. New York: Van Nostrand Reinhold.

Ivanovic, M. (2008) *Cultural Tourism*. Juta: Lansdowne.

Jones, B.T.B. and Murphree, M.W. (2004) Community-based Natural Resource Management as a conservation mechanism: Lessons and directions. In B. Child (ed.) *Parks in Transition* (pp. 63–103). London: Earthscan.

Jones, S. (2005) Community-based tourism: The significance of social capital. *Annals of Tourism Research* 32 (2), 303–324.

Kieti, D. and Akama, J. (2005) Wildlife safari tourism and sustainable local community development in Kenya: A case study of Samburu National Reserve. *Journal of Hospitality and Tourism* 3 (2), 71–81.

Lew, A. (1989) Authenticity and sense of place in tourism development experience of older retail districts. *Journal of Travel Research* 28 (4), 15–22.

Lowenthal, D. (1985) *The Past is a Foreign Country*. Cambridge: Cambridge University Press.

Macnaughten, P. and Urry, J. (1998) *Contested Natures*. London: Sage.

Manwa, H. (2007) Is Zimbabwe ready to venture into the culture tourism market? *Development Southern Africa* 24 (3), 465–474.

Martin, B.S. and Uysal, M. (1990) An examination of the relationship between carrying capacity and the tourism lifecycle: Management and policy implications. *Journal of Environmental Management* 31, 327–333.

Mbaiwa, J.E. (2005) Enclave tourism and its socio-economic impacts in the Okavango Delta, Botswana. *Tourism Management* 26 (2), 157–172.

Michler, I. (2008) A plan for the Okavango. *Africa Geographic*, September 2008, 26.

Mowforth, M. and Munt, I. (1998) *Tourism and Sustainability: A New Tourism in the Third World*. London: Routledge.

Novelli, M. and Gebhardt, K. (2007) Community based tourism in Namibia: 'Reality show' or 'window dressing'? *Current Issues in Tourism* 10 (5), 443–479.

Pattullo, P. (1996) *Last Resorts: The Cost of Tourism in the Caribbean*. London: Wellington House.

Prentice, R. (1994) Heritage: A key sector of the "new tourism". *Progress in Tourism, Recreation and Hospitality Management* 5, 309–324.

Ravenscroft, N. (1999) Hyper-reality in the official (re)construction of leisure sites: The case of rambling. In D. Crouch (ed.) *Leisure/Tourism Geographies: Practices and Geographical Knowledge* (pp. 74–90). London: Routledge.

Redcliffe, M. (1987) *Sustainable Development: Exploring the Contradictions*. London: Methuen.

Saarinen, J. (2006) Traditions of sustainability in tourism studies. *Annals of Tourism Research* 33 (4), 1121–1140.

Saarinen, J. (2007) Cultural tourism, local communities and representations of authenticity: The case of Lesedi and Swazi cultural villages in Southern Africa. In B. Wishetemi, A. Spenley and H. Wels (eds) *Culture and Community: Tourism Studies in Eastern and Southern Africa*. Amsterdam: Rozenberg.

Saarinen, J. and Niskala, M. (2009) Selling places, constructing local cultures in tourism – the role of Ovahimbas in Namibian tourism promotion. In P. Hottola (ed.) *Tourism Strategies and Local Responses in Southern Africa*. Wallingford: CABI Publishing (forthcoming).

Sack, R.D. (1992) *Place, Modernity and Consumer's World*. Baltimore: The John Hopkins University Press.

Scheyvens, R. (2002) *Tourism for Development: Empowering Communities*. Harlow: Prentice-Hall.

Sharpley, R. (2000) Tourism and sustainable development: Exploring the theoretical divide. *Journal of Sustainable Tourism* 8 (1), 1–19.

Spangenberg, J. (2005) Will the information society be sustainable? Towards criteria and indicators for sustainable knowledge society. *International Journal of Innovation and Sustainable Development* 1 (1), 85–102.

Telfer, D. and Sharpley, R. (2007) *Tourism and Development in the Developing World*. London: Routledge.

Teo, P. (2002) Striking a balance for sustainable tourism: Implications of the discourse on globalisation. *Journal of Sustainable Tourism* 10 (4), 459–474.

Teo, P. and Lim, H.L. (2003) Global and local interactions in tourism. *Annals of Tourism Research* 30 (2), 287–306.

Timothy, D. and Boyd, S. (2003) *Heritage Tourism*. Harlow: Prentice-Hall.

Tosun, C. (2000) Limits to community participation in the tourism development process in developing countries. *Tourism Management* 21 (6), 613–633.

Turco, D.M. (1999) Ya' 'at 'ech: A profile of tourists to Navajo Nation. *The Journal of Tourism Studies* 10 (2), 57–61.

Urry, J. (1990) *The Tourist Gaze: Leisure and Travel in Contemporary Societies*. London: Sage.

Urry, J. and Sheller, M. (2006) The new mobilities paradigm. *Environment and Planning A* 38, 207–226.

van Veuren, E.J. (2001) Transforming cultural villages in the spatial development initiatives of South Africa. *South African Geographical Journal* 83 (2), 137–148.

van Veuren, E.J. (2004) Cultural village tourism in South Africa: Capitalizing on indigenous culture. In C.M. Rogerson and G. Visser (eds) *Tourism and Development*

Iissues in Contemporary South Africa (pp. 139–160). Pretoria: Africa Institute of South Africa.

Waitt, G. (1999) Naturalising the 'primitive': A critique of marketing Australia's indigenous peoples as 'hunter-gatherers'. *Tourism Geographies* 1 (2), 142–163.

Wall, G. (1993) International collaboration in the search for sustainable tourism in Bali, Indonesia. *Journal of Sustainable Tourism* 1 (1), 38–47.

Wall, G. (1996) Is ecotourism sustainable? *Environmental Management* 2 (2), 207–216.

Wels, H. (2004) About romance and reality: Popular European imagery in post-colonial tourism in Southern Africa. In C.M. Hall and H. Tucker (eds) *Tourism and Postcolonialism: Contested Discourses, Identities and Representation* (pp. 76–94). London: Routledge.

WCED (1987) *Our Common Future*. Oxford: Oxford University Press.

Wheeller, B. (1993) Sustaining the ego. *Journal of Sustainable Tourism* 1 (1), 121–129.

Wolfe, R.I. (1983) Recreational travel, the new migration revisited. *Ontario Geography* 19, 103–124.

WTO (1993) *Sustainable Tourism Development: Guide for Local Planners*. Madrid: Word Tourism Organization.

Zhou, W. and Richie, J. (2007) Tourism and poverty alleviation: An integrative research framework. *Current Issues in Tourism* 10 (2–3), 119–143.

Index

accommodation 26, 45, 47, 67, 99, 117-118,
121-132, 140, 145, 215, 226-227, 236, 278
Addo National Park 118, 121
African National Congress 43-44, 50
Ai-Ais Transfrontier Park 156, 169, 173,
177-178
alternative tourism 5-7, 23, 27-28, 32, 38,
82-83, 210-211, 255, 271
apartheid 64, 93, 157, 232, 246
– post-apartheid 29, 118
attractions 5-6, 11, 22-23, 27, 46, 51, 55, 73,
81, 101-102, 139-141, 143-144, 148, 150,
161, 195, 199-200, 205, 228, 255-261, 264,
282
authenticity 11, 45, 78, 101, 141, 161, 205,
254-258, 265, 280, 283

backpackers 28
black (economic) empowerment 4, 154
borders (cross-border tourism) 34, 49, 55-56,
117, 142, 170-174, 176, 179, 183-184
business tourism 28

CAMPFIRE 190-193
carrying capacity (tourism) 9, 25-26, 78-82,
87, 177-178, 235
Chobe National Park 144, 174, 178, 195
codes of conduct 5, 273
colonialism 65, 232, 271
commonages 231-232, 237-242, 244, 249-250,
279
communal lands/areas 84-85, 93-98, 105,
107-113, 134, 150-166, 170-171, 174-176,
190-191, 232-235, 278, 282
community involvement/participation
46-48, 82-85, 87, 97, 110, 120-122, 124,
154,163-165, 172, 182-183, 189-191,
194-199, 203-205, 215-216, 223-224,
228-229, 233-236, 277-278, 280-281, 283
community-based natural resource
management 66-67, 70-71, 95, 112,
136-138, 150, 159-165, 169, 172-174,
183-193, 215-220
community-based tourism 4, 7, 37, 47-48,
66-67, 82-85, 94-95, 97-98, 106-113, 146,
150, 154-155, 159-165, 194, 218-222, 271,
279
conservancies 95, 97-98, 106-113, 153-154,
157-165, 173-176, 183, 193, 232, 240,
246-249
conservation 46-47, 49, 82-85, 93-96, 104-109,
112, 116-121, 129-131, 136-142, 147-148,
150-166, 169-173, 176-184, 189-193, 197,
204, 211-212, 234-236, 249-250, 272, 275,
278, 281-283
constrains (in tourism) 20, 35, 37-38, 261
consumption (tourism) 6, 95, 101, 110, 148,
205, 224, 254, 270-271
Cradle of Humankind 101-103
crafts/handicrafts 64-65, 67, 71-73, 83,
161-162, 194-195, 198-202, 204-205,
216-217, 222-223, 237, 248, 273, 283
culinary 257-260, 263-265
cultural tourism 38, 103, 161, 189, 204-205,
222, 283
cultural village 161-162, 282-283

Democratic Alliance 50
degradation 45, 165-166, 171, 233, 240, 275
demonstration effect 224, 228, 279
dependency 37, 204-205, 279
destination 4-6, 8, 11-12, 20-23, 25-29, 32, 34,
36, 38, 45-46, 50, 55, 63-64, 73, 79-81, 86,
99, 104, 112-113, 122, 129-131, 141, 143,
155-158, 165, 210-212, 220-223, 225-227,
253-266, 269-271, 275, 280-284
displacement 158, 180-183, 279
domestic tourism 28-29, 33, 35, 47, 112
Durban 35

ecotourism 23, 37-38, 52, 54, 57, 67, 71, 100,
136, 138-141, 147, 176-179, 184, 210,
233-236, 244, 264, 272
education 4, 6, 12-13, 27, 46, 50, 62, 67, 83,
93, 95, 101, 103, 105, 110, 113, 118, 131,
135, 146, 158, 164, 193, 195, 199, 205, 241,
265, 273
employment 11, 43, 47, 50, 57-58, 61-64,
67-69, 71, 83-84, 93, 98, 111, 147, 174,
194-195, 198, 202-204, 214-223, 227-228,
236, 246, 250, 265, 282-283
empowerment 4, 43, 61, 63-64, 68, 70, 72-73,
85, 87, 110, 119, 146-148, 152, 154, 156,
161, 163-165, 191, 204-205, 216, 275-276,
278-280, 282-283
enclave tourism 26-27, 35, 222-223, 228, 279

entrepreneurship 46-47, 50-51, 104, 111-112, 196, 237, 257, 262, 266, 272-273
ethnic/ethnicity 64, 73, 110, 158, 197, 203, 205, 222, 280
Etosha National Park 106-107, 121-123, 155-156, 175, 177, 271
equity 8, 61-62, 94, 97, 211-212, 216, 228, 278-279 events 28, 55, 58, 99, 155, 259-261
evolution model 81
exotic/exotisation 64-65, 210

Gaborone 28, 46, 139, 214, 225
game reserves 23, 134-138, 155, 157, 162, 174, 195, 199, 282
gender 61-64, 68-69, 275, 278, 282
geotourism 136, 147
globalisation 6-7, 254, 280-281
global change 282
governance 5, 11, 42, 49, 52-53, 94-95, 111-112, 150, 164, 171-172, 278
Great Limpopo Transfrontier Park 172-173, 175-176, 179-183

Harare 24, 214
health 36, 49, 62, 110, 131, 158, 197-198, 264, 273, 275
heritage 11, 47, 63, 65, 72, 83, 93, 100, 101-103, 105, 107, 118, 120, 138-139, 141, 143, 146-148, 189, 199, 222, 236, 253, 257, 259, 264-265, 273, 282-283
HIV/AIDS 35, 65, 226, 276
holism/holistic 8, 86, 106, 155, 160, 270, 278

identity 57, 71, 83, 97, 104, 181-182, 233, 255
income 11, 28, 34, 45, 72, 95, 110-111, 116, 118, 130-131, 157, 160-166, 174, 195, 198, 201, 214, 218, 222, 225, 238, 240-242, 248, 250, 258, 265, 272
inequality 62, 212, 223
Inkatha Freedom Party 50
integrative development/planning 270
Internet 103
investments 3, 5, 44, 46, 48-49, 54-54, 66, 99, 104, 113, 120, 131, 164, 223, 249
Isimangaliso Wetland Park 233-236, 277

Johannesburg 21, 34-35, 214
joint venture 71, 84, 111, 137, 160, 162-163, 194, 199, 203, 217, 249, 280

Kgalagadi National Park 118
Kgalagadi Transfrontier Park 174-175, 178, 183, 195, 199, 202
Kruger National Park 79-80, 116-132, 175, 178, 180, 271, 278-279
KwaZulu-Natal 71, 233, 236

land reform 86, 164, 231-250, 279, 281

leakages 35, 47, 67
life cycle 81, 104
literacy 14, 164, 195, 197, 199, 203
localities 13, 280, 282-283
lodges 6, 67, 71, 119, 157, 161-163, 165, 179, 183, 210, 213-218, 220-221, 225, 228

Makgadikgadi National Park 195
Makgadikgadi Salt Pan 83-84, 140, 144
Maputo 35
mass tourism 5-6, 33, 38, 77, 81, 141, 195
Maun 47, 210, 213-215, 217, 224-225
Millennium Development Goals 4, 21, 63, 146, 274-276
mobility 49, 53-54, 270
Moremi Game Reserve 195, 220-221, 224
motivation (tourist) 5, 24, 101, 121-122, 258

NACOBTA 84, 95, 161-162
Nama Karoo 156
Namaqualand 231-250
Namib Desert 23, 100, 144, 176
Namib-Naukluft National Park 155
Nata sanctuary 83-84
nature-based tourism 47, 81, 136, 143, 155, 173, 176, 179, 184, 226, 244
nature reserves 119, 121, 140, 165, 173-174
new tourism 6-7

Okavango Delta 23, 35, 47, 66, 71, 136, 139, 147, 210-229, 271-272, 275, 277, 279

partnership 4, 51, 62-63, 97, 109, 119, 162, 169, 179, 194, 199, 236, 263, 265, 272, 275-276, 281, 283
Peace Parks 54
poverty alleviation/reduction 4, 11, 21, 34-35, 37, 57, 61-63, 99, 110, 112, 131, 152, 158, 162, 165-166, 183, 189-190, 195-196, 212, 215-216, 218, 228, 271, 274, 275, 281
pro-poor tourism 4, 37-38, 52, 63, 82, 95, 98, 112
prostitution 225-226, 228, 279

Quad-biking 144-146

Reed dance 64-66
resort 6, 23-27, 38, 122, 254
responsible tourism 146, 169, 272
RETOSA 23, 69
Richtersveld National Park 175
rural areas 3, 36, 64, 152, 169, 194-195, 218
– rural communities 70, 101, 153, 159-160, 166, 193-194, 215, 222, 225
– rural tourism 23, 97, 194, 198

safaris 22-24, 27, 46-47, 100, 140, 157, 190, 192, 194-195, 199, 201, 203-205, 213-214, 216-217, 221-222, 224, 226, 228, 279

sanctuary 83-84
second homes 13
sex tourism 64, 225
SMEs 62-64, 66, 72-73, 164
social capital 87, 279-280
South African Communist Party 50
sustainable development 7-11, 13, 53, 78,
    86-87, 99, 135, 151-152, 165, 182, 211-212,
    215-216, 231, 234, 238-239, 270-272,
    275-278, 280-282, 284
sustainable tourism 4-5, 9-11, 13, 37, 65,
    77-87, 93-94, 146, 151, 164, 166, 211-216,
    218, 220-221, 223-224, 227-228, 232,
    244-245, 250, 253-254, 261, 263, 269-272,
    276-281, 283-284

tour operators 25, 162, 211, 223-224, 228,
    235, 260
tourism policy 4, 42-58, 62, 84, 139, 147, 152,
    154, 158, 194, 213, 215, 228, 281
traditions 3, 11, 15, 47, 64-65, 67, 71, 73, 78,
    141, 159-161, 173, 176, 180, 189-190,
    199-201, 205, 221-223, 240, 244, 246, 260,
    279, 282, 283

transformation 6, 28, 93-95, 98, 104, 139, 146,
    150, 154, 278, 281
Transfrontier Parks 169-184, 277
transportation 213, 222, 280-281
trickle down 148

urban 109-110, 134-135, 140, 145, 151, 225
– urban tourism 13, 28

Victoria Falls 23-27, 55, 73, 214-271
village tourism 189-190, 199, 204

White Paper of Tourism 3, 153, 234,
    244-245
wilderness 45-48, 80, 139-140, 143, 147,
    194-195, 199, 212
– wilderness tourism 45-46, 48
wildlife tourism 135, 157, 194, 281
Windhoek 99, 107, 110, 214
wine tourism 260, 262-264
World Bank 21, 44, 53-54
World Heritage Site 101-102, 234-236

zoning 80, 164, 235